Nataliya Metla

The SQP method for optimal control problems with mixed constraints

Nataliya Metla

The SQP method for optimal control problems with mixed constraints

Optimal control of nonlinear elliptic PDEs

Südwestdeutscher Verlag für Hochschulschriften

Impressum/Imprint (nur für Deutschland/ only for Germany)
Bibliografische Information der Deutschen Nationalbibliothek: Die Deutsche Nationalbibliothek verzeichnet diese Publikation in der Deutschen Nationalbibliografie; detaillierte bibliografische Daten sind im Internet über http://dnb.d-nb.de abrufbar.
Alle in diesem Buch genannten Marken und Produktnamen unterliegen warenzeichen-, marken- oder patentrechtlichem Schutz bzw. sind Warenzeichen oder eingetragene Warenzeichen der jeweiligen Inhaber. Die Wiedergabe von Marken, Produktnamen, Gebrauchsnamen, Handelsnamen, Warenbezeichnungen u.s.w. in diesem Werk berechtigt auch ohne besondere Kennzeichnung nicht zu der Annahme, dass solche Namen im Sinne der Warenzeichen- und Markenschutzgesetzgebung als frei zu betrachten wären und daher von jedermann benutzt werden dürften.

Verlag: Südwestdeutscher Verlag für Hochschulschriften Aktiengesellschaft & Co. KG
Dudweiler Landstr. 99, 66123 Saarbrücken, Deutschland
Telefon +49 681 37 20 271-1, Telefax +49 681 37 20 271-0, Email: info@svh-verlag.de
Zugl.: Linz, JKU, Diss., 2008

Herstellung in Deutschland:
Schaltungsdienst Lange o.H.G., Berlin
Books on Demand GmbH, Norderstedt
Reha GmbH, Saarbrücken
Amazon Distribution GmbH, Leipzig
ISBN: 978-3-8381-0227-6

Imprint (only for USA, GB)
Bibliographic information published by the Deutsche Nationalbibliothek: The Deutsche Nationalbibliothek lists this publication in the Deutsche Nationalbibliografie; detailed bibliographic data are available in the Internet at http://dnb.d-nb.de.
Any brand names and product names mentioned in this book are subject to trademark, brand or patent protection and are trademarks or registered trademarks of their respective holders. The use of brand names, product names, common names, trade names, product descriptions etc. even without a particular marking in this works is in no way to be construed to mean that such names may be regarded as unrestricted in respect of trademark and brand protection legislation and could thus be used by anyone.

Publisher:
Südwestdeutscher Verlag für Hochschulschriften Aktiengesellschaft & Co. KG
Dudweiler Landstr. 99, 66123 Saarbrücken, Germany
Phone +49 681 37 20 271-1, Fax +49 681 37 20 271-0, Email: info@svh-verlag.de

Copyright © 2009 by the author and Südwestdeutscher Verlag für Hochschulschriften Aktiengesellschaft & Co. KG and licensors
All rights reserved. Saarbrücken 2009

Printed in the U.S.A.
Printed in the U.K. by (see last page)
ISBN: 978-3-8381-0227-6

"Mathematics is not a deductive science – that's cliché. When you try to prove a theorem, you don't just list the hypotheses, and then start to reason. What you do is trial-and-error, experimentation, and guess work."

Paul Halmos

Abstract

Many scientific and technical processes are described by partial differential equations. The optimization of such processes leads to optimal control problems for partial differential equations. Typically, nonlinear functions and constraints, when some quantities of the process are required to fulfill certain equations and inequalities, are involved in real-life problems. Focus of interest in present work is a family of optimal control problems governed by semilinear elliptic partial differential equations (PDEs) and pointwise nonlinear inequality constraints. In order to find an optimal solution, one puts special attention to numerical methods.

In the scope of present dissertation, we establish necessary and sufficient optimality conditions and analyze the convergence of sequential quadratic programming (SQP) methods applied to mixed constrained optimal control problems, i.e., for the optimal control problem with coupling between control and state in constraints.

The convergence theory for the SQP method bases on its relation to the Newton method applied to a so-called generalized equation which represents first-order necessary optimality conditions.

The necessary optimality conditions will be derived according to the Lagrange technique, which turns to the discussion of the existence of the regular Lagrange multipliers. The uniqueness of Lagrange multipliers is necessary requirement in the convergence analysis. However, the inequality constraints may be active simultaneously, then the uniqueness of associated multipliers are violated. A remedy is to introduce security sets, which contain the active sets at the optimal solution. When these security sets are disjoint the uniqueness takes place.

Sufficient optimality conditions ensure stability under perturbations of the solutions of investigated optimal control problems. Moreover, they present the key for proving the convergence of fast and efficient numerical methods. Until now, sufficient optimality conditions, stability results, and convergence analysis for SQP methods are known in case the pointwise inequality constraints affect solely the control variable.

At the end of this thesis the developed theory is verified by numerical tests for discrete optimal control problems.

Kurzfassung

Eine große Anzahl von naturwissenschaftlichen und technologischen Prozessen wird durch partielle Differentialgleichungen beschrieben. Die Optimierung solcher Prozesse führt auf Optimalsteuerungsprobleme bei partiellen Differentialgleichungen. Reale Anwendungsprobleme sind durch das Auftreten von nichtlinearen Funktionen gekennzeichnet. Außerdem müssen einige der für den Prozess relevanten Größen in gewissen Toleranzbereichen liegen. Im Fokus dieser Arbeit liegt das semilineare (nichtlineare) elliptische Optimalsteuerungsproblem mit punktweisen Ungleichungsnebenbedingungen. Besonderes Augenmerk wird dabei auf die numerische Lösungsmethode gelegt.

Der Schwerpunk der Dissertation ist die Konvergenzanalyse des SQP-Verfahrens (sequential quadratic programming) für gemischt beschränkte Optimalsteuerungsprobleme, d.h. die Ungleichungsnebenbedingungen enthalten sowohl die Steuerung als auch die Prozessgrößen. Außerdem werden die Existenz von Lösungen linearer und semilinearer partielle Differentialgleichungen, sowie notwendige und hinreichende Optimalitätsbedingungen diskutiert.

Die Konvergenztheorie für das SQP-Verfahren wird mit Hilfe der Newton Methode für die sogenannte verallgemeinerte Gleichung entwickelt, dabei sind die notwendigen Optimalitätsbedingungen erster Ordnung in dieser verallgemeinerte Gleichung vertreten.

Notwendige Optimalitätsbedingungen werden mittels der Lagrange-Technik bestimmt, wobei die Existenz von regulären Lagrange-Multiplikatoren benötigt wird. Die Eindeutigkeit der Lagrange-Multiplikatoren ist die notwendige Forderung der Konvergenztheorie. Sind hingegen die gemischten Beschränkung-en gleichzeitig aktiv, dann ist diese Forderung nicht erfüllbar. Als Gegenmittel werden die Sicherheitsmengen, die die in der Lösung aktiven Mengen enthalten, eingeführt. Wenn diese Mengen disjunkt sind, ist die Eindeutigkeit der Lagrange-Multiplikatoren garantiert.

Hinreichende Optimalitätsbedingungen sichern die Stabilität der Lösung des Steuerproblems gegenüber Störungen in den Daten. Außerdem bilden sie den Schlüssel für die Konvergenzanalyse schneller und effizienter numerischer Verfahren. Bis jetzt waren Resultate zu Optimalitätsbedingungen, zur Stabilität und zur Konvergenz von schnellen numerischen Verfahren nur für solche Probleme bekannt, bei denen ausschließlich die Steuerung in den Ungleichungsnebenbedingungen auftritt.

Zum Abschluß wird die entwickelte Theorie durch numerische Ergebnisse für diskretisierte Optimalsteuerungsprobleme bestätigt.

Acknowledgments

This work has been carred out at the Johannes Radon Institute for Computational and Applied Mathematics (RICAM) during the years 2005-2008. I am thankful to Professor Heinz W. Engl, the director of the institute, for arranging an excellent working environment. I have learned a lot from different fields of mathematics.

My first deeply heartfelt thanks go to Prof. Arnd Rösch and Prof. Roland Griesse, adviser and instructors of present work, for proposing me the subject of the dissertation, supervising and inspiring my work, for countless time-intensive discussions, for teaching me how to think and write anew and for reading the manuscript with a great accuracy and making invaluable comments.

At the same time I am greatly indebted to Prof. Helmut Gfrerrer for co-refereeing this work and Prof. Andreas Neubauer who presided to be jury.

Several people had helped me during the project. I owe my thanks to each and every one. Many thanks go to Prof. Walter Alt for helping me to understand the SQP algorithm and for offering the idea with separation of security sets. The members of my working group include Svetlana Cherednichenko, Klaus Krumbieger, Prof. Boris Vexler deserve my gratitude for numerous discussion according to numerical implementation. Thanks to Marie-Therese Wolfram and Mariya Zhariy, who were writing their own dissertations when I was writing mine, have lent a sympathetic ear to groaning, and unwaveringly cheered me on. And thanks to all colleagues of RICAM, particularly Dr. Jenny Niebsch, Dr. Ester Klann for an inspiring research atmosphere and being always warmly welcomed me for limitless questions. Special thanks go to Dr. Stefan Müller and his mother for offering me encouragement and parental care throughout the entire process of drafting the thesis. Among my treasured Berlin friends, Valeria Lykina, Nataliya Togobytska, both read sections of the manuscript, and I benefited enormously from their advices. Thanks also to Dr. Dylan Copeland for proof-reading this thesis, his comments significantly improved the presentation.

My sincere thanks go to my good friend, Alena Furmanchuk, for her appearing in hard moment of desperation and helping me think about and revise present work.

Fortune must have been responsible for involving me to Linz and introducing to me my newest friends. Sincere gratitude goes to all my friends for creation of a lovely social atmosphere, for being every time for me and for providing energy from different places. They have helped me to stay sane through the years of research. Their support and care helped my overcome setbacks and stay focused on my graduate study. I greatly value their friendship and deeply appreciate their belief in me.

A very special section is owed to my family. Of course my parents deserve an especial dedication. Thanks for their love and support, for inspiring me patience and love to mathematics. To my wonderful brother for his amazing energy, savvy, enthusiasm, optimism and love, for being always on my side when I needed some help. I will never be grateful enough for all they have done for me.

Financially support by Austrian Science Fund "Fonds für Förderung der wissenschaftlichen Forschung in Österreich" (FWF) thought the project P18056-N12 is acknowledged.

Finally, I am greatly indebted to them, who are reading these lines, thus all hours spent by writing were justified.

*Dedicated to my mother,
professional mathematician born to be teacher,
on occasion of her 50th birthday anniversary.*

Contents

1 Introduction   **1**

2 Motivation and examples   **5**
    2.1 Boundary control problem 6
    2.2 Distributed control problem 8

3 Finite dimensional optimization problems   **9**
    3.1 Optimality conditions 10
    3.2 SQP algorithm 13

4 Optimal control problem   **17**
    4.1 First-order necessary optimality conditions 19
         4.1.1 Lagrange approach 20
         4.1.2 Regularity of multipliers 21
         4.1.3 Uniqueness of multipliers 24
    4.2 Second-order sufficient optimality condition 26
    4.3 Distributed control problem 31

5 Stability of linear-quadratic problems   **35**
    5.1 Stability result for auxiliary problem 37
    5.2 Relationship between solutions 49
    5.3 Distributed control problem 51

6 Convergence analysis   **53**
    6.1 Sequential quadratic programming method 53
    6.2 Generalized equation 54
    6.3 Local convergence result 56
    6.4 Distributed control problem 65

7 Numerical realization   **67**
    7.1 Primal-Dual Active Set Strategy 67
    7.2 Discrete SQP algorithm 71
    7.3 Numerical tests 78
         7.3.1 An example of boundary optimal control problems . . . 78

       7.3.2   An example of distributed optimal control problems . . . 81

8 Additional Material — 85
   8.1   Convexity 86
   8.2   Bilinear forms 87
   8.3   Elements of measure theory 88
   8.4   Function spaces 89
   8.5   Embedding and trace operator 91
   8.6   Nemyckii-operator 93
   8.7   Elliptic PDEs with Neumann boundary condition 95
   8.8   Regularity of the solution 97
   8.9   Dirichlet boundary problem 99

9 Assumptions — 101

1 Introduction

A host of problems in the natural sciences lead to *optimization*, i.e., to find values of the variables (certain characteristics of the system) that optimize (*minimize* or *maximize*) the *objective* (or *cost function*), that could be profit, time, potential energy, or any quantity or combination of quantities.

A wide variety of technical processes are described by partial differential equations (PDEs), for instance, heat conduction, diffusion, oscillations, electromagnetic waves, flows and other physical phenomena. These problems arise in different areas of applied mathematics, physics, medicine, engineering, including fluid dynamics, optics, solid mechanics, plasma physics, electromagnetic and quantum field theory. The optimization of such processes or identification of material parameters leads to *optimal control problems for PDEs*. The problems in reality are often restricted, i.e., some quantities of the process have to fulfill certain equations and inequalities. It leads to *constrained* optimal control problems. In present work we consider a family of optimal control problems with pointwise nonlinear inequality constraints governed by semilinear elliptic PDEs. We allow for distributed and boundary controls. Problems with mixed control-state constraints present an additional interest as Lavrentiev-type regularization of pointwise state-constrained problems, since the solutions of such problems exhibit better regularity properties than those with purely pointwise state constraints, see [38, 40].

First-order necessary optimality conditions help to find critical points, while sufficient second-order conditions ensure the local optimality of such points. Moreover, they represent the key to proving convergence of fast and efficient numerical methods.

The main subject of this thesis is to establish a local quadratic convergence of the sequential quadratic programming (SQP) method applied to PDE-constrained optimal control problems subject to nonlinear mixed inequality constraints. The special feature of the SQP method is its relationship with Newton's method. Thanks to this, SQP methods have proved to be fast solution methods for nonlinear programming problems. A large body of literature exists concerning the analysis of these methods for finite-dimensional problems, e.g. [4, 34, 42].

1 Introduction

In the classical SQP method (or *Lagrange-Newton method*), developed for problems with equality-type constraints, Newton's procedure is applied to the first-order optimality system, which has the form of a system of equations. In the case of inequality-type constraints, the first-order optimality system cannot be expressed as an equation. However, it can be expressed as an inclusion, or so-called *generalized equation* [36]. It was shown by S.M. Robinson [36] that the Newton procedure applied to such generalized equation is locally quadratically convergent to the solution, if the property called *strong regularity* is satisfied. This approach has been successfully applied to a class of nonlinear cone-constrained optimization problems in infinite-dimensional spaces in [3], and optimal control problems governed by ordinary differential equation (ODE) and subject to the control and/or state constraints, see [6, 7]. Untill now, SQP methods for the optimal control of PDEs are established only for purely control-constrained problems [25, 44, 47].

Following this idea, we exploit the equivalence between the SQP method and a *generalized Newton method*, i.e., Newton's method, applied to a generalized equation representing necessary optimality conditions. We concentrate on specific issues arising due to the presented semilinear elliptic PDE, e.g., careful choice of suitable function spaces. An important step is the verification of the strong regularity of the generalized equation, which is made difficult by the simultaneously present control and state variables in the constraints. The strong regularity means Lipschitz continuous dependence of the solution of the linearized generalized equation on a perturbation parameter. In the context of PDE-constrained optimization, the linearized generalized equation represents necessary and sufficient optimality conditions of a linear-quadratic optimal control subproblem. This interplay between Newton's method and the SQP method is a specific feature, which cannot be derived from general results in Banach spaces [3], since we have to discuss pointwise relations.

In the following chapters we focus on the Lipschitz stability result and local quadratic convergence of the SQP method applied to mixed constrained nonlinear optimal control problems. We derive the optimality conditions following the Lagrange technique, which requires the introduction of so-called *dual* variables (or *(Lagrange) multipliers*). We also show the existence of the regular multipliers. At the end, the expected quadratic behavior of the SQP method applied to model problems will be illustrated on numerical examples. The practical implementation is based on the *discretization* concept, see e.g. [11, 15, 41]. The discretization of the optimality system leads to solving a high-dimensional system of equations and offers discrete optimal solution of the nonlinear problem. The convergence result means in this case the convergence of the corresponding discrete values. In order to find the solution of the optimization problem involving inequality constraints, we apply the primal-dual active sets (PDAS) algorithm [9].

Thesis outline

In **Chapter 2** we describe some motivating examples for the class of optimal control problems to be studied in following chapters.

A brief outline of the analysis of the finite-dimensional optimization problems is given in **Chapter 3**. We formulate the optimality condition and recall the general procedure of the SQP method for these problems. The analysis carried out for optimal control problem governed by PDE and mixed constraints is based on the general ideas presented in Chapter 3. It stands to reason to consider such simple type of problems, where one does not need to take care about complicated spaces, to get clear understanding of the analysis in subsequence.

In **Chapter 4** we specify the exact problem setting. We also study the necessary and sufficient optimality conditions for the optimal control problems we deal with. The Lagrange approach shows to be a good technique for this aim, but to use it we need to make sure of the existence of associated dual variables (or so-called Lagrange multipliers). This important step is carried out in Chapter 4 as well.

Chapter 5 deals with linear-quadratic optimal control problems. We prove stability results for the solutions of these subproblems. This stability estimate is the main ingredient in convergence analysis of the SQP method for the solution of the nonlinear optimal control problem, studied in the previous chapter.

The main result on local quadratic convergence of the SQP method for the problems we study is realized in **Chapter 6**. Here we define generalized equation and point out the link between the generalized equation and first-order necessary optimality conditions for nonlinear mixed constrained optimal control problems governed by elliptic PDE. Based on the interplay between the optimality system and generalized equation, the generalized Newton method yields a local quadratic convergence result for the SQP method.

Finally, the numerical verification of the developed theory is presented in **Chapter 7**. The problem solving via computer leads to the corresponding discrete version of involved quantities and equations. Thus, for practical implementation of the SQP method we consider its discretization. The primal-dual active set strategy is applied to deal with inequality constraints in each

1 Introduction

iteration step. Chapter 7 deals with the discretization technique and tools of PDAS method. At the end some numerical tests are given.

Appendix 8 contains some mathematical results from the functional analysis and analysis of partial differential equations as well as optimization theory that are used frequently in the text.

For the convenience of readers the list of all assumptions is collected in **Appendix 9**.

2 Motivation and examples

In this chapter few examples will be used to represent the variety of domains where one finds optimization problems considered in the present work. The optimal control problems constrained by elliptic PDE appear by modeling physical phenomena such as cooling or heating processes with controlling of energy expenses, diffusions and flows, weather forecasting to predict the state of atmosphere, or in biochemistry, to determine the geometry of a molecule, where various techniques are possible (X-ray crystallography, nuclear magnetic resonance).

Let us consider stationary heating problems, see e.g. [46].

The equilibrium distribution of absolute temperature $y : \Omega \to \mathbb{R}^+$ inside a body (see pic. 2.0.1) is determined by stationary heat equation

$$-\operatorname{div}(\kappa \nabla y) = f \quad \text{in } \Omega, \qquad (2.0.1)$$

where κ is the body's thermal conduction, and $f : \Omega \to \mathbb{R}_0^+$ presents possible heat sources. In the simple situation κ is a positive constant, but, in general, it can depend on both y and the space coordinate $\xi \in \Omega$. To complete the problem description for the determination of the equilibrium temperature distribution in Ω, one needs to formulate boundary conditions on Γ. The appropriate choice of boundary condition depends on the physical situation to be modeled as well as on what quantity to be physically measured and controlled in the situation of the interest. If the temperature on Γ is known, one deals

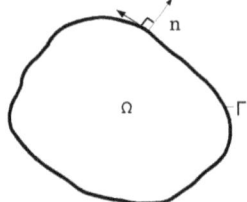

Figure 2.0.1: Motivating example

2 Motivation and examples

with the *Dirichlet* boundary condition, i.e.,

$$y = y_0 \quad \text{on } \Gamma, \tag{2.0.2}$$

where $y_0 : \Gamma \to \mathbb{R}^+$ is the known temperature on the boundary. For example, if the interior of some apparatus is heated in a room temperature environment, then, by choosing Ω sufficiently large, one can ensure that y_0 is known to be the room temperature. In a different situation, the temperature distribution y_0 on Γ might also be unknown, if it is controlled by means of a heating device. Then one deals with the *Neumann* boundary condition, i.e.,

$$\partial_n y = \alpha(y_0 - y) \quad \text{on } \Gamma, \tag{2.0.3}$$

where n is an outer normal vector associated with the element ds on surface Γ and α is a heat transmission number from Ω to the surrounding medium.

The general idea of an *optimal control problem* associated with such processes is to vary (or *control*) an input quantity (called *control* and denoted by u) such that some output quantity (called *state* and denoted by y) has a desired property. This desired property is measured according to some function J (called *objective* or *cost functional*). Usually, the objective is formulated such that the desired optimal case coincides with the minimum (or maximum) of J. In general, J can depend on both the control u and the state y. However, if there exists a unique state for each control, then the objective can be considered as a function of the control alone.

By controlling partial differential equations (PDEs), the state is a quantity determined as a solution of corresponding PDE, whereas the control can be an input function prescribed on the boundary Γ (so-called *boundary control*) or an input function prescribed on the domain Ω (so-called *distributed control*).

2.1 Boundary control problem

Consider the case, where the optimal temperature distribution $y_\Omega : \Omega \to \mathbb{R}^+$ is known and heating elements can control the temperature $u := y_0$ at each point on the boundary Γ. The goal is to find u such that the actual temperature y approximates the desired y_Ω. This problem leads to minimization of the

2.1 Boundary control problem

objective functional

$$J(y,u) := \frac{1}{2}\int_\Omega (y(\xi)-y_\Omega(\xi))^2 d\xi + \frac{\gamma}{2}\int_\Gamma u(\xi)^2 ds(\xi)$$
$$= \frac{1}{2}\|y-y_\Omega\|^2_{L^2(\Omega)} + \frac{\gamma}{2}\|u\|^2_{L^2(\Gamma)},$$

where $\gamma > 0$ and s is used to denote the surface measure on Γ. The second integral in the definition of J is a typical companion of the first one in this type of problem. For instance, it can be interpreted as measuring the energy costs of the heating device. From the mathematical point of view, the second term of J has a *regularizing effect* (called *Tikhonov regularization*). It counteracts the tendency of the control to become locally unbounded and rugged as J approaches its infimum.

Due to physical and technical limitations of the heating device, one needs to impose some restrictions on the control and state. Physical limitations result from the fact that any device will be destroyed, if its temperature becomes too high or too low. Usually the technical limitations of the heating device will be much more restrictive, providing upper and lower bounds for the temperatures that the device can impose. Hence, such limitations lead to the *control, state* or *mixed control-state* constraints. In more general situations, the optimal control problem involves nonlinear functions.

To find a good approximate solution for the desired mathematical model, one has to rely on the numerical technique, since only simple cases can be solved analytically. In order to validate the numerical results, it is necessary to know convergence properties of the method applied to the initial problem. Typically, one solves nonlinear constrained optimization problem using various Newton type methods applied to its system of optimality conditions. The SQP (or generalized Newton) method proved to be a fast and efficient approximation algorithm. That is why we focus on the convergence analysis of the mentioned numerical method for PDE constrained optimal control problems.

In the next chapters, we will deal with the following *boundary control problem*:

$$\text{Minimize} \quad J(y,u) := \int_\Omega \phi(\xi,y(\xi))\,d\xi + \int_\Gamma \psi(\xi,y(\xi),u(\xi))\,ds(\xi) \quad \textbf{(P)}$$

subject to $u \in L^\infty(\Gamma)$, and nonlinear elliptic boundary state equation

$$\begin{aligned} \mathcal{A}y + d(\xi,y) &= f \quad \text{in } \Omega, \\ \partial_n y + b(\xi,y) &= u \quad \text{on } \Gamma \end{aligned} \qquad (2.1.1)$$

2 Motivation and examples

as well as pointwise nonlinear mixed boundary constraints

$$g_i(\xi, y(\xi), u(\xi)) \leqslant 0 \text{ a.e. on } \Gamma, \ i = 1, ..., s \ (s \in \mathbb{N}) \quad (2.1.2)$$

where $\mathcal{A}$ is an elliptic operator satisfying **(A1)**. All necessary assumptions will be introduced later in the text, where it is firstly needed. For convenient reading we also list all used assumptions in Appendix 9.

2.2 Distributed control problem

Now we consider the case where we can control the heat source f inside the domain Ω, i.e., we set $u := f$ in (2.0.1). The control is no longer concentrated on the boundary Γ but *distributed* over Ω. Such distributed heat sources occur, for instance, by electromagnetic or microwave heating. Since u lives in Ω, the corresponding integration in the cost function has to be performed over Ω. Moreover, the constraints have to be imposed in Ω too.

In the sequel, we consider a more general situation of *distributed control problem*:

$$\text{Minimize} \quad J(y, u) := \int_\Omega \phi(\xi, y(\xi), u(\xi)) \, d\xi \quad (\mathbf{P'})$$

subject to $u \in L^\infty(\Omega)$, and nonlinear elliptic boundary state equation

$$\begin{aligned} \mathcal{A}y + d(\xi, y) &= u \quad \text{in } \Omega, \\ y &= 0 \quad \text{on } \Gamma \end{aligned} \quad (2.2.1)$$

as well as pointwise nonlinear mixed constraints

$$g_i(\xi, y(\xi), u(\xi)) \leqslant 0 \text{ a.e. in } \Omega, \ i = 1, ..., s. \quad (2.2.2)$$

Remark 2.2.1. *In the sequel, all analysis is carried out on the boundary control problem* **(P)**. *At the end of each chapter the reader can find the corresponding results concerning the distributed control problem* **(P')**. *However, the proofs for the latter problems are omitted.*

3 Finite dimensional optimization problems

We begin our study of nonlinear optimal control for PDEs, described in the previous chapter, by discussing constrained optimization problems defined on a finite dimensional space. The tast is to find the minimum of a function $J : \mathbb{R}^n \to \mathbb{R}$ subject to constraints on the variables:

$$\min_{x \in \mathcal{M}} J(x), \qquad (3.0.1)$$

where $\mathcal{M} := \{x \in \mathbb{R}^n \mid c_i(x) = 0, \; i \in \mathbf{E}; \; c_i(x) \leqslant 0, \; i \in \mathbf{I}\}$, and $\mathbf{E}$ and $\mathbf{I}$ are two separated finite sets of indices. Here c_i are all smooth and real-valued functions on a subset of $\mathbb{R}^n$. The reason for studying problems on finite dimensional space separately is that they allow us to see the principles for constructing local approximation algorithms (in particular, SQP and Newton's methods). That also will be the basis of our algorithms for the problems stated on function spaces, without requiring any knowledge of Lebesgue and Sobolev spaces or analysis of PDEs. The algorithms for nonlinear control problems governed by PDEs and pointwise inequality constraints will be more tangled than those in present chapter. However, understanding of the basic approach here should help in more intricate cases.

The theory presented in this chapter can be found in following books for optimization [4, 16, 34]. We start with a definition of the optimal solution for (3.0.1).

Definition 3.0.2.

(i) A point x is called an <u>admissible (feasible) point</u> if $x \in \mathcal{M}$, and $\mathcal{M} \subset \mathbb{R}^n$ is called an <u>admissible (feasible) set</u>.

(ii) A point x^* is called a <u>local optimal solution</u> if there exists a neighborhood $U(x^*)$ of x^* such that the inequality $J(x^*) \leqslant J(x)$ holds for all $x \in U(x^*) \cap \mathcal{M}$.

(iii) A local optimal solution according to (ii) is called <u>strict</u> if $J(x^*) < J(x)$ holds for all admissible $x \in U(x^*)$, $x \neq x^*$.

3 Finite dimensional optimization problems

The studying of optimization problem includes the answers to the questions:

(1) Does the problem (3.0.1) have an optimal solution?

(2) Which conditions does an optimal solution have to satisfy?

(3) By means of which numerical method can one determine an optimal solution?

In following we focus on the second and third questions.

In one-dimension calculus, by studying of the differentiable function $J : \mathbb{R} \to \mathbb{R}$, one learns that vanishing of the derivative $J'(x^*) = 0$ is a *necessary* condition for J to have an extremum (minimum or maximum) at x^*. The simple example (e.g. $J(x) = x^3$ at $x^* = 0$) shows that this necessary condition is not sufficient for J to have an extremum (minimum or maximum) at x^*.

Similarly, we formulate necessary optimality conditions that are of first order in the sense that they involve only the first derivatives in multiple finite dimensions for (3.0.1) in the next section and even in infinite-dimensional spaces for optimal control problems of PDEs in Section 4.1.

Remark 3.0.3. *For convex optimal problems the necessary optimality conditions are also sufficient.*

3.1 Optimality conditions

To solve the constrained optimization problems one introduces an auxiliary function (called *Lagrange function* or *Lagrangian*) that depends on the *primal variable* x and *dual variable* λ (called *Lagrange multiplier*) associated to the constraints. Then one presents necessary optimality conditions of the first order in terms of the gradients of the Lagrange function with respect to both variables x and λ.

The Lagrangian for (3.0.1) is defined as

$$\mathcal{L}(x,\lambda) = J(x) + \sum_{i \in \mathbf{E} \cup \mathbf{I}} \lambda_i c_i(x), \qquad (3.1.1)$$

where the vector λ consists of components λ_i, $i \in \mathbf{E} \cup \mathbf{I}$. One can prove the existence of the dual variables λ_i only under certain *regularity conditions* (or so-called *constraint qualifications*). There exist various forms of the regularity conditions. In practice weaker constraint qualifications are preferred since they provide stronger optimality conditions. In the present analysis we are using so-called *linearized Slater condition* (or *local Slater condition*).

3.1 Optimality conditions

Definition 3.1.1. *Let x^* be a local solution of* (3.0.1). *We say that* (3.0.1) *satisfies the <u>linearized Slater condition</u> if there exist $\tau > 0$ and $\tilde{x} \in \mathbb{R}^n$ such that*

$$c_i(x^*) + \nabla c_i(x^*)\tilde{x} \leqslant -\tau, \quad i \in \mathbf{I}$$
$$\nabla c_i(x^*)\tilde{x} = 0, \quad i \in \mathbf{E}.$$

Theorem 3.1.2. *Suppose that x^* is a local solution of* (3.0.1) *and the linearized Slater condition holds at x^*. Then there is a Lagrange multiplier vector λ^*, with components λ_i^*, $i \in \mathbf{E} \cup \mathbf{I}$, such that the following conditions are valid*

$$\nabla_x \mathcal{L}(x^*, \lambda^*) = 0$$
$$c_i(x^*) = 0, \quad i \in \mathbf{E}, \tag{3.1.2}$$
$$\lambda_i^* \geqslant 0, \quad c_i(x^*) \leqslant 0, \quad \lambda_i^* c_i(x^*) = 0, \quad i \in \mathbf{I},$$

If the gradients of the active inequality constraints and the gradients of the equality constraints are linearly independent at x^*, then the Lagrange multiplier vector λ^* is unique.

Proof. See [4, Theorem 7.2.18]. □

The system of first-order optimality conditions (3.1.2) is also called *Karush-Kuhn-Tucker conditions* (*KKT conditions*). While the first-order necessary conditions (3.1.2) assume that x^* is a local solution and deduce properties of J and c_i, the *second-order sufficient condition* (**SSC**, see Assumption 3.1.3 below) is a condition on J and c_i, $i \in \mathbf{E} \cup \mathbf{I}$, that ensures that x^* is a local optimal solution of the problem (3.0.1), see [34, Section 12.5].

Assumption 3.1.3. *Suppose that for some admissible point $x^* \in \mathbb{R}^n$ there is a Lagrange multiplier vector λ^* such that the KKT conditions* (3.1.2) *are satisfied and*

$$\delta x^\top \nabla_{xx} \mathcal{L}(x^*, \lambda^*) \delta x > 0 \tag{3.1.3}$$

holds for all nonzero $\delta x \in \mathbb{R}^n$ satisfying

$$\begin{cases} \nabla c_i(x^*)\delta x = 0, & i \in \mathbf{E}, \\ \nabla c_i(x^*)\delta x = 0 \text{ and } \lambda_i^* > 0, & i \in \mathcal{A}(x^*), \\ \nabla c_i(x^*)\delta x \leqslant 0 \text{ and } \lambda_i^* = 0, & i \in \mathcal{A}(x^*), \end{cases} \tag{3.1.4}$$

where $\mathcal{A}(x^*) := \{i \in \mathbf{I} \mid c_i(x^*) = 0\}$ is called the active set at x^*.

3 Finite dimensional optimization problems

Remark 3.1.4. *The condition (3.1.3) is equivalent to existence of $\alpha > 0$ such that*

$$\delta x^\top \nabla_{xx} \mathcal{L}(x^*, \lambda^*)\, \delta x \geqslant \alpha |\delta x|^2 \qquad (3.1.5)$$

for all elements $\delta x \in \mathbb{R}^n$ satisfying (3.1.4). (Note that this inequality holds trivially for $\delta x = 0$.)

Theorem 3.1.5. *Suppose that Assumption 3.1.3 holds. Then there exist $\beta > 0$ and $\varepsilon > 0$ such that*

$$f(x) \geqslant f(x^*) + \beta |x - x^*|^2 \qquad (3.1.6)$$

holds for any $x \in B_\varepsilon(x^*) \cap \mathcal{M}$, $x \neq x^*$, i.e., x^* is a strict local optimal solution for (3.0.1).

Proof. see [34, Theorem 12.6] or [4, Theorem 7.3.1]. □

Note (3.1.6) is known as *second-order growth condition*.

View on optimal control for PDEs

In main aspects the study of the optimal control for PDE is similar to the finite dimensional case. However, the set $\mathcal{M}$ will be defined on a functional space X and contain an elliptic equation. Here is an additional discussion about the existence of the solution of the present PDE, regularity of this solution and careful choice of the corresponding functional space appear (see Section 8.7). In order to state the optimality conditions for the optimal control problems subject to PDE and pointwise inequality constraints, the Lagrange approach provides a good technique again. However, the study of existence of regular Lagrange multipliers associated to PDE and pointwise inequality constraints becomes more interesting (see Section 4.1). In contrast to the finite dimension case, for uniqueness of the Lagrange multipliers here we will require separation of security sets, see Definition 4.1.5 and Assumption (**A6**). Moreover, in practice the form (3.1.5) of **SSC** is not always applicable. The reason is so-called *two-norm discrepancy* [1].

Remark 3.1.6. *While the norm on X has to be chosen sufficiently strong (usually of L^∞-type) in order to guarantee twice differentiability of the nonlinear PDE (and sometimes of the objective), the coercivity estimate*

$$\mathcal{L}_{xx}(x^*, \lambda^*)(\delta x, \delta x) \geqslant \alpha \, \|\delta x\|_{\tilde{X}}^2$$

---
[1] One estimates the element from the space X using the norm in space $\tilde{X}$.

is only verified for a weaker norm $\|\cdot\|_{\tilde{X}}$ (usually L^2-type).

This observation motivates a modified form of **SSC** in the sequel (see Section 4.2). In this aspect, one needs to look for a *strict local optimal solution in the sense of X* (see Definition 4.0.2). In particular, we will show that also for PDE-constrained control problem **SSC** yields the second-order growth condition (4.0.3) in context of two-norm discrepancy principle and so x^* is a strict local optimal solution (see Theorem 4.2.5).

3.2 SQP algorithm

The local SQP method for the finite dimensional optimization problem (3.0.1) is discussed in present section. Its basic idea is analogous to Newton's method for unconstrained minimization: at each step, a local model of the optimization problem is constrained and solved, yielding a step toward the solution of the original problem. An SQP method uses a quadratic model for the objective function and a linear model for the constraints. A nonlinear problem in which the objective function is quadratic and the constraints are linear is called *linear-quadratic* or *quadratic* problem, abbreviate by **LQP** or **QP**, respectively. An SQP method solves a QP in each iteration.

Thus the SQP algorithm replaces (3.0.1) with

$$
\begin{aligned}
\text{Minimize} \quad & \nabla J(x^k)(x-x^k) + \frac{1}{2}(x-x^k)^\top \nabla_{xx}\mathcal{L}(x^k,\lambda^k)(x-x^k) \\
\text{subject to} \quad & c_i(x^k) + \nabla c_i(x^k)(x-x^k) = 0, \quad i \in \mathbf{E}, \\
& c_i(x^k) + \nabla c_i(x^k)(x-x^k) \leqslant 0, \quad i \in \mathbf{I}.
\end{aligned}
\tag{3.2.1}
$$

The local convergence property of the SQP method is well understood when (x^*, λ^*) satisfies Assumption 3.1.3. If starting point x^0 is sufficiently close to x^*, and the Lagrange multiplier $\{\lambda^k\}$ remain sufficiently close to λ^*, then the sequence $\{x^{k+1}\}$ converges to x^* at a second-order rate, where x^{k+1} solves (3.2.1).

The strategy for solving (3.2.1) is to make the decision about which of the inequality-constraints appear to be active at the solution x^k internally during the solution of the quadratic problem x^{k+1}. This variant explicitly maintains working active sets $\mathcal{A}^k := \{i \in \mathbf{I} \mid c_i(x^k) + \nabla c_i(x^k)(x-x^k) = 0\}$ of apparently

3 Finite dimensional optimization problems

active indices and solves the quadratic programming problem

$$\text{Minimize} \quad \nabla J(x^k)(x - x^k) + \frac{1}{2}(x - x^k)^\top \nabla_{xx}\mathcal{L}(x^k, \lambda^k)(x - x^k) \qquad (3.2.2)$$
$$\text{subject to} \quad c_i(x^k) + \nabla c_i(x^k)(x - x^k) = 0, \quad i \in \mathcal{A}^k$$

to find the step x^{k+1}. The contents of $\mathcal{A}^k$ are updated at each iteration by examining the Lagrange multipliers for the subproblem (3.2.2) and examining the values of $c(x^{k+1})$ at the iterate x^{k+1} for $i \notin \mathcal{A}^k$. The advantage of the convex finite dimensional optimization problem is that the active sets $\mathcal{A}^k$ in each iterate coincide with $\mathcal{A}(x^*)$, if (x^k, λ^k) is sufficiently close to (x^*, λ^*), see Theorem 3.2.1 below. Therefore, by solving (3.0.1), one can restrict oneself to the solution of the corresponding equality constrained optimization problem, where the inequalities are substituted by the tallying equalities.

Theorem 3.2.1. *(see [34, Theorem 18.1])*
Suppose that x^* is a solution of (3.0.1) *at which the KKT conditions* (3.1.2) *are satisfied for some λ^*. Suppose, too, that the gradients of the active inequality constraints and the gradients of the equality constraints are linearly independent at x^*, Assumption 3.1.3 is valid and strict complementarity holds, i.e., $\lambda_i^* > 0$ for each $i \in \mathcal{A}(x^*)$. Then if (x^k, λ^k) is sufficiently close to (x^*, λ^*), there is a local solution of the subproblem* (3.2.1) *whose active set $\mathcal{A}^k$ is the same as the active set $\mathcal{A}(x^*)$ of the nonlinear problem* (3.0.1) *at x^*.*

Let us consider now the equality-constrained optimization problem, that is,

$$\text{Minimize} \quad J(x)$$
$$\text{subject to} \quad c_i(x) = 0, \quad i \in \mathbf{E}. \qquad (3.2.3)$$

If x^* is a local minimizer of (3.2.3), the linearized Slater condition at x^* holds, and $\lambda^* = (\lambda_i^*)_{i \in \mathbf{E}}$ is associated Lagrange multiplier vector, then the first-order optimality conditions

$$F(x^*, \lambda^*) := \begin{pmatrix} \nabla_x \mathcal{L}(x^*, \lambda^*) \\ \nabla_\lambda \mathcal{L}(x^*, \lambda^*) \end{pmatrix} = 0 \qquad (3.2.4)$$

hold. Thanks to Assumption 3.1.3, $\nabla_{xx}\mathcal{L}(x^k, \lambda^k)$ is positive definite whenever x^k is close to x^*. Moreover, if the gradients of the active inequality constraints and the gradients of the equality constraints are linearly independent at x^*, then $\nabla c(x^k)$ has full row rank thanks to Theorem 3.2.1. Therefore, the Jaco-

3.2 SQP algorithm

bian

$$\begin{pmatrix} \nabla_{xx}\mathcal{L}(x^k,\lambda^k) & \nabla^\top c(x^k) \\ \nabla c(x^k) & 0 \end{pmatrix}$$

is non-singular for any x^k in the vicinity of x^*. So it is reasonable to approximate (x^*, λ^*) by applying Newton's approach to (3.2.4), i.e., for given (x^k, λ^k) the Newton method defines the next iterate as the solution of the linear equation

$$F(x^k,\lambda^k) + \begin{pmatrix} \nabla_x F(x^k,\lambda^k) \\ \nabla_\lambda F(x^k,\lambda^k) \end{pmatrix} \cdot \begin{pmatrix} x - x^k \\ \lambda - \lambda^k \end{pmatrix}$$

$$= \begin{pmatrix} \nabla_x \mathcal{L}(x^k,\lambda^k) \\ \nabla_\lambda \mathcal{L}(x^k,\lambda^k) \end{pmatrix} + \begin{pmatrix} \nabla_{xx}\mathcal{L}(x^k,\lambda^k) & \nabla^\top c(x^k) \\ \nabla c(x^k) & 0 \end{pmatrix} \cdot \begin{pmatrix} x - x^k \\ \lambda - \lambda^k \end{pmatrix} = 0.$$

(3.2.5)

On other hand, assuming that (x^k, λ^k) is sufficiently close to (x^*, λ^*) and that Assumption 3.1.3 is satisfied, the quadratic problem

$$\text{Minimize} \quad \nabla J(x^k)(x - x^k) + \frac{1}{2}(x - x^k)^\top \nabla_{xx}\mathcal{L}(x^k, \lambda^k)(x - x^k)$$

$$\text{subject to} \quad c_i(x^k) + \nabla c_i(x^k)(x - x^k) = 0, \quad i \in \mathbf{E}$$

is convex and therefore possesses the unique solution (x^{k+1}, λ^{k+1}) which fulfills the optimality system (3.2.5).

That means, in other words the SQP method is equivalent to Newton's method applied to the first-order optimality conditions (3.2.4). Therefore, at least locally, the SQP method defines a good step from (x^k, λ^k), which converges to (x^*, λ^*) quadratically, see [16, Theorem 2.4.3].

Finally, turning back to the inequality-constrained problem (3.0.1) we recall Theorem 3.2.1. Under assumptions of Theorem 3.2.1 the active set $\mathcal{A}^k$ remains fixed, so the subproblem (3.2.1) behaves like the corresponding equality-constrained quadratic program, i.e., we can eventually ignore the inequality constraints that do not fall into the active set $\mathcal{A}(x^*)$, while treating the active constraints as equality constraints, [34]. Therefore, the SQP method provides a local quadratic convergence of iterates to the optimal solution also for the inequality-constrained problem (3.0.1).

3 Finite dimensional optimization problems

View on optimal control for PDEs

The analysis of the SQP method for nonlinear optimal control problems will involve technical details due to the present PDE as well as the pointwise mixed inequality constraints. Similarly, we are going to construct the convergence analysis with the help of Newton's approach. However, due to the pointwise relation of the inequality constraints, we are not able to achieve a result similar to Theorem 3.2.1. It means that, in contrast to the finite dimensional case discussed above, we cannot ignore the role of inequality constraints, since the active sets can change from one iterate to another even close to the soltion. Considering the optimal control problems governed by PDEs and pointwise inequality constraints, we will introduce a so-called *set-valued* function corresponding to the inequality constraints, and following the technique described above we will pose first-order necessary conditions as an equation, or more precisely, an inclusion (called *generalized equation*) which will contain the set-valued function (see Section 6.2). The proof of local quadratic convergence of the SQP method for such problems will be based on the perturbation arguments and the verification of certain property (called *strong regularity*) of the generalized equation (see Section 6.3).

4 Optimal control problem

In this chapter we introduce a family of nonlinear optimal control problems governed by semilinear PDEs and pointwise inequality constraints. One discerns distributed and boundary control problems governed by nonlinear or/and linear mixed inequality constraints. We will restrict all of the following analysis on the more involved *boundary control problems* governed by finite number of nonlinear constraints, that is,

$$\text{Minimize} \quad J(y,u) := \int_\Omega \phi(\xi, y(\xi))\, d\xi + \int_\Gamma \psi(\xi, y(\xi), u(\xi))\, ds(\xi) \quad \text{(P)}$$

subject to $u \in L^\infty(\Gamma)$, and nonlinear elliptic boundary state equation

$$\begin{aligned} \mathcal{A} y + d(\xi, y) &= f \quad \text{in } \Omega, \\ \partial_n y + b(\xi, y) &= u \quad \text{on } \Gamma \end{aligned} \quad (4.0.1)$$

as well as pointwise nonlinear mixed boundary constraints

$$g_i(\xi, y(\xi), u(\xi)) \leqslant 0 \text{ a.e. on } \Gamma, \ i = 1, ..., s. \quad (4.0.2)$$

In the following, we assume that Ω is a bounded domain in $\mathbb{R}^N$, $N = \{2, 3\}$, which has a $C^{1,1}$ boundary Γ. In analysis of elliptic PDEs (for details see Appendix 8.7) in order to prove the existence of a solution for (4.0.1) one supposes the following conditions to hold.

Assumption.

(A1) The operator $\mathcal{A}: H^1(\Omega) \to H^1(\Omega)^*$ is defined as $\mathcal{A}y(v) = a[y, v]$, where

$$a[y, v] = ((\nabla v)^\top, A_0 \nabla y) + (a_0 y, v).$$

A_0 is an $N \times N$ symmetric matrix with Lipschitz continuous entries on $\overline{\Omega}$ such that $\rho^\top A_0(\xi) \rho \geqslant m_0 |\rho|^2$ holds with some $m_0 > 0$ for all $\rho \in \mathbb{R}^N$ and almost all $\xi \in \overline{\Omega}$. Moreover, $a_0 \in L^\infty(\Omega)$. The symbol ∂_n denotes the co-normal derivative associated to A_0.

4 Optimal control problem

The bilinear form $a[\cdot,\cdot]$ is assumed to be continuous and coercive, i.e.,

$$a[y,v] \leq \bar{c}\|y\|_{H^1(\Omega)}\|v\|_{H^1(\Omega)},$$
$$a[y,y] \geq \underline{c}\|y\|_{H^1(\Omega)}^2$$

for all $y, v \in H^1(\Omega)$ with some positive constants $\bar{c}$ and $\underline{c}$. (This is satisfied if ess inf $a_0 > 0$.) The right hand side f is taken from $L^N(\Omega)$.

(A2) The functions $d(\xi, y)$ and $b(\xi, y)$ belong to the class of C^2 with respect to y for almost all $\xi \in \Omega$ or $\xi \in \Gamma$, respectively. Moreover, d_{yy} and b_{yy} are assumed be locally bounded and locally Lipschitz-continuous functions with respect to y, i.e., the following conditions hold true:
there exist $K_d > 0$ and $K_b > 0$ such that

$$|d(\xi, 0)| + |d_y(\xi, 0)| + |d_{yy}(\xi, 0)| \leq K_d,$$
$$|b(\xi, 0)| + |b_y(\xi, 0)| + |b_{yy}(\xi, 0)| \leq K_b,$$

and for any $M > 0$, there exist $L_d(M) > 0$ and $L_b(M) > 0$ such that

$$|d_{yy}(\xi, y_1) - d_{yy}(\xi, y_2)| \leq L_d(M)|y_1 - y_2| \quad \text{a.e. in } \Omega,$$
$$|b_{yy}(\xi, y_1) - b_{yy}(\xi, y_2)| \leq L_b(M)|y_1 - y_2| \quad \text{a.e. on } \Gamma$$

for all $y_1, y_2 \in \mathbb{R}$ satisfying $|y_1|, |y_2| \leq M$.

Additionally for all $y \in \mathbb{R}$ we assume $d_y(\xi, y) \geq 0$ a.e. in Ω and $b_y(\xi, y) \geq 0$ for almost all $\xi \in \Gamma$.

Due to Lemma 8.8.6, under these assumptions the elliptic equation (4.0.1) possesses a unique solution $y \in W^{1,p}(\Omega)$, $p \in [1, \infty)$ for every $u \in L^\infty(\Gamma)$. For the further discussion, we fix $\bar{p} \in (N, \infty)$ and define

$$X := W^{1,\bar{p}(\Omega)} \times L^\infty(\Gamma).$$

Note that $W^{1,\bar{p}}(\Omega) \hookrightarrow C(\overline{\Omega})$ owing to Sobolev's embedding Theorem 8.5.4. From now on, we assume that $x^* := (y^*, u^*) \in X$ is a local optimal solution of **(P)**. We will turn our attention on a *strict local optimal solution* in the following sense.

Definition 4.0.2. *A point $x^* \in X$ is called a <u>strict local optimal solution in the sense of L^∞</u> if there exists $\varepsilon > 0$ such that the inequality $J(x^*) < J(x)$ holds for all admissible $x \in X \setminus \{x^*\}$ with $\|x - x^*\|_{L^\infty(\Omega) \times L^\infty(\Gamma)} \leq \varepsilon$.*

In this chapter we discuss first and second order optimality conditions for **(P)** in the form of Lagrange multipliers. Moreover, we examine so-called *second* order growth condition (see Definition 4.0.3 below), which guarantees x^* to be the strict local minimizer for (**P**) according to Definition 4.0.2, see Theorem 4.2.5 in the sequel.

Definition 4.0.3. *The second-order growth condition holds, if there exists $\varepsilon > 0$ and $\beta > 0$ such that*

$$J(x) \geqslant J(x^*) + \beta \|x - x^*\|^2_{L^2(\Omega) \times L^2(\Gamma)} \qquad (4.0.3)$$

is valid for all $x := (y, u) \in X$ satisfying $\|x - x^*\|_{L^\infty(\Omega) \times L^\infty(\Gamma)} \leqslant \varepsilon$.

4.1 First-order necessary optimality conditions

As we have seen in the previous chapter, one of central problems in the optimization theory is to derive optimality conditions. They play an outstanding role in such aspects as regularity properties of optimal pairs and numerical methods. By analogy with finite dimensional case, we employ the Lagrange approach.

Assumption.

(A3) *The function $\psi(\xi, y, u)$ is measurable with respect to $\xi \in \Gamma$ for each y and u, and of class C^2 with respect to y and u for almost all $\xi \in \Gamma$. The second derivatives are assumed to be locally bounded and locally Lipschitz-continuous functions, i.e., the following conditions hold:
there exists $K_\psi > 0$ such that*

$$|\psi(\xi, 0, 0)| + |\psi_u(\xi, 0, 0)| + |\psi_y(\xi, 0, 0)|$$
$$+ |\psi_{uu}(\xi, 0, 0)| + |\psi_{yu}(\xi, 0, 0)| + |\psi_{yy}(\xi, 0, 0)| \leqslant K_\psi$$

and for any $M > 0$, there exists $L_\psi(M) > 0$ such that

$$|\psi_{yy}(\xi, y_1, u_1) - \psi_{yy}(\xi, y_2, u_2)| \leqslant L_\psi(M)(|y_1 - y_2| + |u_1 - u_2|),$$
$$|\psi_{yu}(\xi, y_1, u_1) - \psi_{yu}(\xi, y_2, u_2)| \leqslant L_\psi(M)(|y_1 - y_2| + |u_1 - u_2|),$$
$$|\psi_{uy}(\xi, y_1, u_1) - \psi_{uy}(\xi, y_2, u_2)| \leqslant L_\psi(M)(|y_1 - y_2| + |u_1 - u_2|),$$
$$|\psi_{uu}(\xi, y_1, u_1) - \psi_{uu}(\xi, y_2, u_2)| \leqslant L_\psi(M)(|y_1 - y_2| + |u_1 - u_2|)$$

for all $y_i, u_i \in \mathbb{R}$ satisfying $|y_i|, |u_i| \leqslant M$, $i = 1, 2$. Analogous conditions are assumed to hold for $g_i = g_i(\xi, y, u)$, $i = 1, ..., s$ and $\phi = \phi(\xi, y)$.

Similar to the finite dimensional case, see Section 3.1, in order to prove the existence of Lagrange multipliers associated to (**P**) we assume to hold the

4 Optimal control problem

linearized Slater condition, which in view of problem setting (**P**) is given by Assumption (**A4**) below.

Assumption.

(**A4**) *There exist $\tau > 0$ and $\hat{u} \in L^\infty(\Gamma)$ such that*

$$g_i(y^*, u^*) + g_{i,y}(y^*, u^*)\hat{y} + g_{i,u}(y^*, u^*)\hat{u} \leqslant -\tau$$

holds a.e. on Γ, where $\hat{y} \in W^{1,\bar{p}}(\Omega)$ is the unique solution of the linearized PDE

$$\mathcal{A}\hat{y} + d_y(y^*)\hat{y} = 0 \quad \text{in } \Omega,$$
$$\partial_n \hat{y} + b_y(y^*)\hat{y} = \hat{u} \quad \text{on } \Gamma.$$

4.1.1 Lagrange approach

Following the technique from Section 3.1, we define the Lagrange functional with multipliers in corresponding dual spaces [1]

$$\mathcal{L}: W^{1,\bar{p}}(\Omega) \times L^\infty(\Gamma) \times W^{1,\bar{p}'}(\Omega) \times \left[L^\infty(\Gamma)^*\right]^s \to \mathbb{R}$$

by

$$\mathcal{L}(y, u, p, \mu_1, \ldots, \mu_s) = J(y, u) + a[y, p] + (d(y) - f, p)_\Omega + (b(y) - u, p)_\Gamma$$
$$+ \sum_{i=1}^{s} \langle \mu_i, g_i(y, u) \rangle_{L^\infty(\Gamma)^*, L^\infty(\Gamma)},$$

where $1/\bar{p} + 1/\bar{p}' = 1$. Compared to finite dimensional optimization, the structure of $\mathcal{L}$ appears more complicated. However, under Assumption (**A4**) the following first-order necessary conditions of Karush-Kuhn-Tucker type can be shown.

Lemma 4.1.1. *Under Assumptions* (**A1**)–(**A4**), *there exist non-negative functionals $\mu_i \in L^\infty(\Gamma)^*$, $i = 1, \ldots, s$, and a function $p \in W^{1,q}(\Omega)$, $1 \leqslant q <$*

---

[1] We observe for $y \in W^{1,\bar{p}}(\Omega)$, we have $\mathcal{A}y \in W^{1,\bar{p}'}(\Omega)^*$, where $1/\bar{p} + 1/\bar{p}' = 1$. Therefore $a[y, \cdot] \in W^{1,\bar{p}'}(\Omega)^*$ is well defined for $p \in W^{1,\bar{p}'}(\Omega)$.

$N/(N-1)$ such that the following conditions are satisfied:

$$\begin{cases} A^*p + d_y(y^*)p = -\phi_y(y^*) & \text{in } \Omega \\ \partial_n^* p + b_y(y^*)p = -\psi_y(y^*, u^*) - \sum_{i=1}^{s} g_{i,y}(y^*, u^*)\mu_i & \text{on } \Gamma \end{cases} \quad (4.1.1)$$

$$(\psi_u(y^*, u^*) - p, h)_\Gamma + \sum_{i=1}^{s} \langle g_{i,u}(y^*, u^*)h, \mu_i \rangle_\Gamma = 0 \quad \text{for all } h \in L^\infty(\Gamma) \quad (4.1.2)$$

$$\langle g_i(y^*, u^*), \mu_i \rangle = 0 \quad \text{for } i = 1, \ldots, s. \quad (4.1.3)$$

Note (4.1.1) is to be meant by weak sence.

Proof. The proof can be carried out analogously to [2] or [14]. □

In the terminology of optimal control the optimality condition (4.1.1) is usually referred to as *adjoint state equation* corresponding to the state equation (4.0.1), while relation (4.1.2) is called *gradient equation*. The function p is known as an *adjoint state*.

4.1.2 Regularity of multipliers

In the present form the Lagrange multipliers $\mu_i \in L^\infty(\Gamma)^*$ are hardly perspicuous. Our goal now is to show that the multipliers μ_i, $i = 1, \ldots, s$ and adjoint state p are indeed more regular functions, namely that $\mu_i \in L^\infty(\Gamma)$ and $p \in W^{1,\bar{p}}(\Omega)$. Following [40], we verify first that all μ_i are measurable functions, then that they belong to the space $L^1(\Gamma)$. Finally, using the bootstrapping argument we will obtain the desired regularity.

To this end, we assume the Assumptions **(A1)**–**(A4)** be fulfilled.

Let us show that the Lagrange multipliers are measurable functions. Thanks to Theorem 8.3.7, each $\mu \in [L^\infty(\Gamma)]^*$ for $i = 1, \ldots, s$ can be uniquely presented by the sum

$$\mu = \mu_c + \mu_p,$$

where μ_c is countably additive and μ_p is purely finitely additive. Moreover, if $\mu \geqslant 0$ holds, then μ_c and μ_p are non-negative as well owing to Theorem 8.3.6. Due to Theorem 8.3.5 the purely additive part μ_p can be characterized in the following way:

if λ is a non-negative and countably additive measure on Γ, then there exists a decreasing sequence $\Gamma \supset E_1 \supset E_2 \supset \cdots \supset E_n \supset \cdots$ such that $\lim_{n \to \infty} \lambda(E_n) = 0$ and $\mu_p(E_n) = \mu_p(\Gamma)$ for all n.

4 Optimal control problem

We shall apply this result with the Lebesgue measure λ on Γ in order to show that purely finitely additive part $\mu_{p,i}$ vanishes on Γ, that will imply $\mu_i \in L^1(\Gamma)$ for all $i = 1,...,s$. Since our inequality constraints are equi-directed due to Assumption (**A5**) below, we can simplify the proof given in [17]. From now on, the functions ψ and g_i, $i = 1,\ldots,s$ are assumed to satisfy the following conditions.

Assumption.

(**A5**) *There is a constant $m > 0$ such that the properties*

$$\psi_{uu}(\xi,y,u) \geqslant m \quad a.e. \text{ on } \Gamma, \ \forall (y,u) \in \mathbb{R}^2,$$

$$\text{and} \quad g_{i,u}(\xi,y,u) \geqslant m \quad a.e. \text{ on } \Gamma, \ \forall (y,u) \in \mathbb{R}^2$$

hold.

Lemma 4.1.2. *Suppose that Assumptions (**A1**)–(**A5**) are satisfied. Then the purely finite additive parts of each μ_i vanish, and μ_i belong to $L^1(\Gamma)$ for $i = 1, \ldots, s$.*

Proof. Consider the singular part $\mu_{p,j}$ for a fixed index $j \in \{1, \ldots, s\}$. Thanks to Theorem 8.3.5, there exists a decreasing sequence $\{E_n\} \subset \Gamma$ such that $\lambda(E_n) \to 0$ and

$$\langle \mu_{p,j}, \chi_\Gamma \rangle = \langle \mu_{p,j}, \chi_{E_n} \rangle.$$

We set $h = \chi_{E_n}/m$ in (4.1.2) and obtain

$$-\frac{1}{m}(\psi_u(y^*,u^*)-p, \chi_{E_n})_\Gamma = \frac{1}{m}\sum_{i=1}^{s}\langle g_{i,u}(y^*,u^*)\chi_{E_n}, \mu_i\rangle \geqslant \frac{1}{m}\langle g_{j,u}(y^*,u^*)\chi_{E_n}, \mu_{p,j}\rangle$$

$$\geqslant \langle \chi_{E_n}, \mu_{p,j}\rangle = \langle \chi_\Gamma, \mu_{p,j}\rangle = \|\mu_{p,i}\|_{L^\infty(\Gamma)^*}, \quad \forall\, n.$$

For $n \to \infty$, the left hand side tends to 0, which implies $\mu_{p,j} = 0$. Finally, the Radon-Nikodym Theorem [27, Theorem 2, Section 6.5.3] implies that $\mu_{c,j}$ can be identified with a function in $L^1(\Gamma)$. □

Thanks to Lemma 4.1.2 we can rewrite the gradient equation (4.1.2) in the form

$$\psi_u(y^*,u^*) - p + \sum_{i=1}^{s} g_{i,u}(y^*,u^*)\mu_i = 0 \quad a.e. \text{ in } \Gamma. \tag{4.1.4}$$

Next we show that the regularity of the Lagrange multipliers $\mu_i \in L^1(\Gamma)$ for $i = 1, \ldots, s$ and adjoint state $p \in W^{1,q}(\Omega)$, $1 \leqslant q < N/(N-1)$ can be improved, and that, in fact, by definition of Lagrange functional $\mathcal{L}$ instead of dual spaces $L^\infty(\Gamma)^*$ and $W^{-1,\bar{p}}(\Omega)^*$ we can use $L^\infty(\Gamma)$ and $W^{1,\bar{p}}(\Omega)$ and that $\mathcal{L}$ has a regular structure.

Theorem 4.1.3. *If Assumptions* **(A1)**–**(A5)** *are satisfied, then the Lagrange multipliers μ_i belong to $L^\infty(\Gamma)$, $i = 1, \ldots, s$, and the adjoint state p is an element of $W^{1,\bar{p}}(\Omega)$.*

Proof. We employ the bootstrapping argument using (4.1.1) and (4.1.4). Due to Lemma 4.1.1 the adjoint state p belongs to the space $W^{1,q}(\Omega)$ for $1 \leqslant q < N/(N-1)$. The trace operator maps $W^{1,q}(\Omega)$ to $W^{1-1/q,q}(\Gamma)$, see Theorem 8.5.7. Thanks to Sobolev's embedding Theorem 8.5.4 we find $W^{1-1/q,q}(\Gamma) \hookrightarrow L^r(\Gamma)$ for $r \in [1, (N-1)/(N-2))$. Now (4.1.4) implies that

$$\sum_{i=1}^{s} g_{i,u}(y^*, u^*)\mu_i = p - \psi_u(y^*, u^*) \in L^r(\Gamma),$$

since $\psi_u(y^*, u^*) \in L^\infty(\Gamma)$. Due to Assumption **(A5)**, $g_{i,u}(y^*, u^*) \geqslant m$ holds, and nonnegativity of μ_i implies that $\mu_i \in L^r(\Gamma)$ for all $i = 1, \ldots, s$. Using the fact that $g_{i,y}(y^*, u^*) \in L^\infty(\Gamma)$, $\psi_y(y^*, u^*) \in L^\infty(\Gamma)$ and $\phi_y(y^*) \in L^\infty(\Omega)$, we conclude that the right hand sides in (4.1.1) are elements of $L^\infty(\Omega)$ and $L^r(\Gamma)$, respectively.

From Lemma 8.8.1 we impose that $p \in W^{1,r}(\Omega)$ for $r \in [1, (N-1)/(N-2))$. In case $N = 2$, we find that $r < \infty$ and $p \in C(\overline{\Omega})$ by continuous embedding. Using (4.1.2) we obtain that $\mu_i \in L^\infty(\Gamma)$. In case $N = 3$, we repeat the procedure above and find $\mu_i \in L^{\tilde{r}}(\Gamma)$ with $\tilde{r} < 4$. Again Lemma 8.8.1 provides $p \in W^{1,\tilde{r}}(\Omega)$, which is continuously embedded into $C(\overline{\Omega})$ for $r > 3$, see Theorem 8.5.4, thus $\mu_i \in L^\infty(\Gamma)$ holds.

The assertion $p \in W^{1,\bar{p}}(\Omega)$ follows from Lemma 8.8.1 because the right hand sides of adjoint state equation (4.1.1) are in $L^\infty(\Omega)$ and $L^\infty(\Gamma)$, which clearly embeds into $L^{\bar{p}}(\Omega)$ and $L^{\bar{p}}(\Gamma)$, respectively. □

Thanks to these intermediate results we can simplify the structure of $\mathcal{L}$ and redefine the Lagrange functional

$$\mathcal{L} : W^{1,\bar{p}}(\Omega) \times L^\infty(\Gamma) \times W^{1,\bar{p}}(\Omega) \times \left[L^\infty(\Gamma)\right]^s \to \mathbb{R}$$

4 Optimal control problem

by

$$\mathcal{L}(y,u,p,\mu_1,\ldots,\mu_s) = J(y,u) + a[y,p] + (p,d(y)-f)_\Omega$$
$$+ (p,b(y)-u)_\Gamma + \sum_{i=1}^s (\mu_i, g_i(y,u))_\Gamma. \quad (4.1.5)$$

Then the system of optimality (4.1.1)–(4.1.3) reads as

$$a[v,p] + (d_y(y^*)p, v)_\Omega + (b_y(y^*)p, v)_\Gamma + (\phi_y(y^*), v)_\Omega$$
$$+ (\psi_y(y^*, u^*), v)_\Gamma + \sum_{i=1}^s (g_{i,y}(y^*, u^*)\mu_i, v)_\Gamma = 0 \quad (4.1.6a)$$
$$\text{for all } v \in W^{1,\bar{p}'}(\Omega)$$

$$\psi_u(y^*, u^*) - p + \sum_{i=1}^s g_{i,u}(y^*, u^*)\mu_i = 0, \quad \text{a.e. on } \Gamma \quad (4.1.6b)$$

$$a[y,v] + (d(y),v)_\Omega + (b(y),v)_\Gamma - (f,v)_\Omega - (u,v)_\Gamma = 0 \quad (4.1.6c)$$
$$\text{for all } v \in W^{1,\bar{p}'}(\Omega)$$

$$0 \leqslant \mu_i \perp g_i(y^*, u^*) \leqslant 0, \quad i=1,\ldots,s, \quad \text{a.e. on } \Gamma \quad (4.1.6d)$$

Note the last line decodes

$$\mu_i \geqslant 0, \quad g_i(y^*, u^*) \leqslant 0, \quad \mu_i g_i(y^*, u^*) = 0 \quad \text{a.e. on } \Gamma \text{ for } i=1,\ldots,s.$$

4.1.3 Uniqueness of multipliers

From (4.1.6a)–(4.1.6d) we observe that if μ_i, $i=1,\ldots,s$ are simultaneously nonzero on a set of positive measure, the multipliers $(p, \mu_1, \ldots, \mu_s)$ may not be unique. The simple example 4.1.4 below confirms this fact. The goal of this section is to find a way to get around such critical situation.

Example 4.1.4. *Let Ω be an arbitrary bounded domain with Lipschitz boundary, ε and γ be positive real numbers.*

$$\text{Minimize } \frac{1}{2}\|y\|^2_{L^2(\Omega)} + \frac{1}{2}\|y\|^2_{L^2(\Gamma)} + \frac{\gamma}{2}\|u - u_d\|^2_{L^2(\Gamma)}$$
$$\text{subject to } \begin{cases} -\Delta y + y = 0 & \text{in } \Omega, \quad u \leqslant 0 \quad \text{on } \Gamma \\ \partial_n y = u & \text{on } \Gamma, \quad \varepsilon u + y \leqslant 0 \quad \text{on } \Gamma. \end{cases}$$

Suppose that $u_d := \gamma^{-1}(\varepsilon + \tau S1)$, where $S1$ is the solution of the PDE for $u \equiv 1$ and τ is a trace operator (see Appendix 8.3). Note $u_d \geqslant 0$ due to the maximum principle (see Corollary 8.7.4), therefore $\|u - u_d\|^2_{L^2(\Gamma)} > \|u_d\|_{L^2(\Gamma)}$.

4.1 First-order necessary optimality conditions

This implies, in particular, that $y = u = 0$ is the unique solution of this problem. Any tuple (p, μ_1, μ_2) satisfying (4.1.6a), (4.1.6b) and (4.1.6d), i.e.,

$$-\Delta p + p = 0 \quad \text{in } \Omega \qquad \mu_1 \geqslant 0, \qquad \mu_2 \geqslant 0 \quad \text{a.e. on } \Gamma$$
$$\partial_n p = -\mu_2 \quad \text{on } \Gamma \qquad -\gamma u_d - p + \mu_1 + \varepsilon \mu_2 = 0 \quad \text{a.e. on } \Gamma$$

is a set of Lagrange multipliers for the problem. It is easy to check that $(p, \mu_1, \mu_2) = (-S1, 0, 1)$ and $(p, \mu_1, \mu_2) = (0, \varepsilon + \tau S1, 0)$ both satisfy this system, and so does any convex combination.

In order to avoid the situation when the multipliers are nonunique, we introduce so-called *security sets* and require Assumption **(A6)** below.

Definition 4.1.5. Let $\sigma_1, \ldots, \sigma_s > 0$. For every $i = 1, \ldots, s$ the <u>security set of level</u> σ_i is defined as $S_i := \{\xi \in \Gamma : -\sigma_i \leqslant g_i(y^*, u^*) \leqslant 0\}$.

Assumption.

(A6) Suppose there exist $\sigma_1, \ldots, \sigma_s > 0$ such that $S_i \cap S_j = \emptyset$ for all $i, j = 1, \ldots, s$, $i \neq j$. Moreover, we assume that the boundary value problem

$$\mathcal{A}^* p + d_y(y^*)p = r_1 \quad \text{in } \Omega,$$
$$\partial^* p + b_y(y^*)p \sum_{i=1}^{s} \chi_{S_i} g_{i,u}^{-1}(y^*, u^*) g_{i,y}(y^*, u^*) = r_2 \quad \text{on } \Gamma \quad (4.1.7)$$

has a unique weak solution $p \in H^1(\Omega)$ for all right hand sides $r_1 \in L^2(\Omega)$ and $r_2 \in L^2(\Gamma)$.

Note **(A6)** means that in a small neighborhood of the optimal pair (y^*, u^*) characterized by σ_i the inequality constraints (4.0.2) are disjointly active.

Theorem 4.1.6. *Let Assumptions* **(A1)**–**(A6)** *be satisfied. Then the Lagrange multipliers and the adjoint state are unique.*

Proof. Let us fix some index $j \in \{1, \ldots, s\}$. We multiply the gradient equation (4.1.6b) by χ_{S_j} and obtain

$$\chi_{S_j} \mu_j = \chi_{S_j} g_{j,u}^{-1}(y^*, u^*) [p - \psi_u(y^*, u^*)] \quad \text{a.e. on } \Gamma. \quad (4.1.8)$$

Owing to **(A6)** $\mu_j = \chi_{S_j} \mu_j$ holds. Plugging this and (4.1.8) into the adjoint equation (4.1.6a), we obtain (4.1.7) with

$$r_1 = -\phi_y(y^*), \qquad r_2 = -\psi_y(y^*, u^*) + \sum_{i=1}^{s} \chi_{S_i} g_{i,u}^{-1}(y^*, u^*) g_{i,y}(y^*, u^*) \psi_u(y^*, u^*).$$

4 Optimal control problem

By Assumption **(A6)**, the adjoint state p is unique. In view of (4.1.8), also the multipliers μ_i are unique. □

Thus, under Assumptions **(A1)**–**(A6)** and if x^* is a optimal solution of **(P)**, there exists the unique tuple $(p^*, \mu_1^*, \ldots, \mu_s^*)$ that together with $x^* = (y^*, u^*)$ satisfies the optimality conditions (4.1.6a)–(4.1.6d).

4.2 Second-order sufficient optimality condition

This section turns our attention to the second-order sufficient optimality condition (**SSC**), that guarantees x^* to be a local minimizer of **(P)**. In fact, we will show that **SSC** yields the second order growth condition.

Let $\lambda^* = (p^*, \mu_1^*, \ldots, \mu_s^*)$ be the Lagrange multiplier vector associated to $x^* = (y^*, u^*)$. We abbreviate $w = (y, u, p, \mu_1, \ldots, \mu_s)$ and set

$$W := W^{1,\bar{p}}(\Omega) \times L^\infty(\Gamma) \times W^{1,\bar{p}}(\Omega) \times [L^\infty(\Gamma)]^s.$$

Owing to Assumptions **(A2)**–**(A3)** the Lagrangian $\mathcal{L}(\cdot)$ is twice continuous differentiable and we find

$$\mathcal{L}_{xx}(w)(\delta x, \delta x) := \int_\Omega \left(\phi_{yy} + d_{yy} p\right)(\delta y)^2 d\xi$$

$$+ \int_\Gamma \begin{pmatrix}\delta y \\ \delta u\end{pmatrix}^\top \begin{pmatrix} \psi_{yy} + b_{yy} p + \sum_{i=1}^s \mu_i g_{i\,yy} & \psi_{yu} + \sum_{i=1}^s \mu_i g_{i\,yu} \\ \psi_{uy} + \sum_{i=1}^s \mu_i g_{i\,uy} & \psi_{uu} + \sum_{i=1}^s \mu_i g_{i\,uu} \end{pmatrix} \begin{pmatrix}\delta y \\ \delta u\end{pmatrix} ds \quad (4.2.1)$$

Let the element w^* satisfy first-order necessary conditions (4.1.6a)–(4.1.6d). The form (4.2.1) is linear with respect to multipliers and locally Lipschitz continuous with respect to x under Assumptions **(A2)**–**(A3)** and Lemma 8.6.4. That is, for sufficient small $\varepsilon > 0$ there exists a positive constant $C_\mathcal{L}$ such that for all $\|x - x^*\|_{L^\infty(\Omega) \times L^\infty(\Gamma)} \leq \varepsilon$ the estimate

$$\left|\left(\mathcal{L}_{xx}(x^* + \theta(x - x^*), \lambda^*) - \mathcal{L}_{xx}(x^*, \lambda^*)\right)(\delta x, \delta x)\right|$$
$$\leq C_\mathcal{L} \|x - x^*\|_{L^\infty(\Omega) \times L^\infty(\Gamma)} \|\delta x\|^2_{L^2(\Omega) \times L^2(\Gamma)} \quad (4.2.2)$$

holds for all $\theta \in (0,1)$. Analogous to the finite dimensional case Assumption **(A7)** below introduces **SSC** for **(P)**.

4.2 Second-order sufficient optimality condition

Assumption.

(A7) *There exists a constant $\alpha > 0$ such that*

$$\mathcal{L}_{xx}(w^*)(\delta x, \delta x) \geqslant \alpha \, \|\delta x\|^2_{L^2(\Omega) \times L^2(\Gamma)} \qquad (4.2.3)$$

for all $\delta x = (\delta y, \delta u) \in X$ satisfying the linearized PDE

$$\begin{aligned} \mathcal{A}\,\delta y + d_y(y^*)\,\delta y &= 0 \quad \text{in } \Omega \\ \partial_n \delta y + b_y(y^*)\,\delta y &= \delta u \quad \text{on } \Gamma. \end{aligned} \qquad (4.2.4)$$

Remark 4.2.1. *There exist weaker sufficient conditions: for real numbers $\delta_i > 0$, $i = 1,...,s$ there exists $\alpha > 0$ such that the inequality (4.2.3) holds for all $\delta x = (\delta y, \delta u) \in X$ satisfying the linearized PDE and*

$$g_i(x^*) + g_{i,x}(x^*)\,\delta x = 0 \quad \text{on } A_i(\delta_i),\ i = 1,...,s,$$

where $A_i(\delta_i) := \{x \in \Gamma : \mu_i \geqslant \delta_i\}$ is a <u>strongly active set</u>. For more details see [37, 39]. However, to avoid any unnecessary technicalities we do not use them in the further analysis.

Note Assumption **(A7)** implies the strong Legendre-Clebsch condition (4.2.5) below.

Lemma 4.2.2. *Suppose that Assumptions **(A1)**–**(A3)** and **(A7)** hold. Then*

$$\psi_{uu}(y^*, u^*) + \sum_{i=1}^{s} g_{i,uu}(y^*, u^*)\mu_i^* \geqslant \alpha \qquad (4.2.5)$$

holds a.e. on Γ.

Proof. See [45, Lemma 5.1]. $\square$

In the next two lemmas we study Assumption **(A7)**. While (4.2.3) is assumed to hold for all $\delta x \in X$ satisfying linearized PDE (4.2.4), Lemma 4.2.3 shows that the similar inequality, (4.2.6) below, is valid also for all $x \in X$, in a neighborhood of the optimal solution x^*, which satisfies semilinear elliptic equation (4.0.1). Moreover, assuming that w^* satisfies Assumption **(A7)** one can show that each $\tilde{w}$ in the L^∞-vicinity of w^* fulfills a similar assumption, see Lemma 4.2.4 below.

Lemma 4.2.3. *There exist $\varepsilon > 0$ and $\alpha' > 0$ such that*

$$\mathcal{L}_{xx}(w^*)(x - x^*, x - x^*) \geqslant \alpha' \, \|x - x^*\|^2_{L^2(\Omega) \times L^2(\Gamma)} \qquad (4.2.6)$$

4 Optimal control problem

for all $x - x^* = (y - y^*, u - u^*) \in X$ provided that $\|x - x^*\|_{L^\infty(\Omega) \times L^\infty(\Gamma)} \leqslant \varepsilon$ with x and x^* satisfying the semilinear state equation (4.0.1).

Proof. Let x satisfy (4.0.1). We define $\delta x = (\delta y, \delta u) \in X$ with $\delta u = u - u^*$ such that (4.2.4) is valid. Then the error $e := y^* - y - \delta y$ fulfills the linear PDE

$$\begin{aligned} \mathcal{A} e + d_y(y^*)\, e &= d(y) - d(y^*) - d_y(y^*)(y - y^*) && \text{in } \Omega \\ \partial_n e + b_y(y^*)\, e &= b(y) - b(y^*) - b_y(y^*)(y - y^*) && \text{on } \Gamma. \end{aligned} \quad (4.2.7)$$

The right hand sides of (4.2.7) we denote by r_1 and r_2, respectively, and estimate by

$$\begin{aligned} \|r_1\|_{L^2(\Omega)} + \|r_2\|_{L^2(\Gamma)} &= \left\| \int_0^1 [d_y(y^* + s(y - y^*)) - d_y(y^*)]\, ds\, (y - y^*) \right\|_{L^2(\Omega)} \\ &\quad + \left\| \int_0^1 [b_y(y^* + s(y - y^*)) - b_y(y^*)]\, ds\, (y - y^*) \right\|_{L^2(\Gamma)} \\ &\leqslant L_d \int_0^1 s\, ds\, \|y - y^*\|_{L^\infty(\Omega)} \|y - y^*\|_{L^2(\Omega)} \\ &\quad + L_b \int_0^1 s\, ds\, \|y - y^*\|_{L^\infty(\Gamma)} \|y - y^*\|_{L^2(\Gamma)}. \end{aligned}$$

Thanks to the triangle inequality, we obtain

$$\begin{aligned} \|r_1\|_{L^2(\Omega)} + \|r_2\|_{L^2(\Gamma)} &\leqslant \frac{L_d}{2} \|y - y^*\|_{L^\infty(\Omega)} \left(\|\delta y\|_{L^2(\Omega)} + \|e\|_{L^2(\Omega)} \right) \\ &\quad + \frac{L_b}{2} \|y - y^*\|_{L^\infty(\Gamma)} \left(\|\delta y\|_{L^2(\Gamma)} + \|e\|_{L^2(\Gamma)} \right). \end{aligned}$$

In view of Lemma 8.6.4, $d_y(y^*) \in L^\infty(\Omega)$ and $b_y(y^*) \in L^\infty(\Gamma)$. The Sobolev embedding Theorem 8.5.4 and Theorem 8.8.4 imply the existence of the unique solution e of (4.2.7) and an a priori estimate

$$\|e\|_{H^1(\Omega)} \leqslant c_\Omega \left(\|r_1\|_{L^2(\Omega)} + \|r_2\|_{L^2(\Gamma)} \right).$$

Using the embedding $H^1(\Omega) \hookrightarrow L^2(\Omega)$ and $H^1(\Omega) \hookrightarrow L^2(\Gamma)$, we obtain

$$\|e\|_{L^2(\Omega)} \leqslant c'\, L\varepsilon \left(\|\delta y\|_{L^2(\Omega)} + \|e\|_{L^2(\Omega)} \right),$$

where $L = \max\{L_d, L_b\}$ and the positive constant c' involves c_Ω and corresponding embedding constants. For sufficiently small $\varepsilon > 0$, we can absorb the last term in the left hand side and obtain

$$\|e\|_{L^2(\Omega)} \leqslant c''(\varepsilon)\, \|\delta y\|_{L^2(\Omega)}$$

where $c''(\varepsilon) = \frac{1}{1-c'L\varepsilon} - 1 \searrow 0$ as $\varepsilon \searrow 0$. A straightforward application of Lemma 8.2.3 concludes the proof. □

Lemma 4.2.4. *There exists $R > 0$ and $\alpha'' > 0$ such that*

$$\mathcal{L}_{xx}(\tilde{w})(x,x) \geq \alpha'' \|x\|^2_{L^2(\Omega) \times L^2(\Gamma)}$$

holds for all $(y,u) \in X$:

$$\mathcal{A} y + d_y(\tilde{y}) \, y = 0 \quad \text{in } \Omega,$$
$$\partial_n y + b_y(\tilde{y}) \, y = u \quad \text{on } \Gamma,$$

provided that $\|\tilde{w} - w^*\|_{W^\infty} < R$ with $W^\infty := L^\infty(\Omega) \times L^\infty(\Gamma) \times L^\infty(\Omega) \times [L^\infty(\Gamma)]^s$.

Proof. Let (y, u) be an arbitrary pair satisfying

$$\mathcal{A} y + d_y(\tilde{y}) \, y = 0 \quad \text{in } \Omega,$$
$$\partial_n y + b_y(\tilde{y}) \, y = u \quad \text{on } \Gamma.$$

We define $\hat{y} \in W^{1,\bar{p}}(\Omega)$ as the solution of

$$\mathcal{A} \hat{y} + d_y(y^*) \, \hat{y} = 0 \quad \text{in } \Omega,$$
$$\partial_n \hat{y} + b_y(y^*) \, \hat{y} = u \quad \text{on } \Gamma,$$

for the same control u as in the previous PDE. Then $\delta y := y - \hat{y}$ satisfies

$$\mathcal{A} \, \delta y + d_y(y^*) \, \delta y = \big(d_y(y^*) - d_y(\tilde{y})\big) y \quad \text{in } \Omega,$$
$$\partial_n \delta y + b_y(y^*) \, \delta y = \big(b_y(y^*) - b_y(\tilde{y})\big) y \quad \text{on } \Gamma.$$

A standard a priori estimate and the triangle inequality yield

$$\begin{aligned}
\|\delta y\|_{H^1(\Omega)} &\leq c_\Omega \, \|d_y(y^*) - d_y(\tilde{y})\|_{L^\infty(\Omega)} \, \|y\|_{L^2(\Omega)} \\
&\quad + c_\Omega \, \|b_y(y^*) - b_y(\tilde{y})\|_{L^\infty(\Gamma)} \, \|y\|_{L^2(\Gamma)} \\
&\leq c_\Omega \, \|d_y(y^*) - d_y(\tilde{y})\|_{L^\infty(\Omega)} \big(\|\hat{y}\|_{L^2(\Omega)} + \|\delta y\|_{L^2(\Omega)}\big) \\
&\quad + c_\Omega \, \|b_y(y^*) - b_y(\tilde{y})\|_{L^\infty(\Gamma)} \big(\|\hat{y}\|_{L^2(\Gamma)} + \|\delta y\|_{L^2(\Gamma)}\big).
\end{aligned}$$

Due to the Lipschitz property of $d_y(\cdot)$ and $b_y(\cdot)$ with respect to L^∞-norm, there exists a function $c(R)$ tending to 0 as $R \to 0$, such that

$$\max\{\|d_y(y^*) - d_y(\tilde{y})\|_{L^\infty(\Omega)}, \|b_y(y^*) - b_y(\tilde{y})\|_{L^\infty(\Gamma)}\} \leq c(R),$$

4 Optimal control problem

provided that $\|\tilde{y} - y^*\|_{L^\infty(\Omega)} < R$ and $\|\tilde{y} - y^*\|_{L^\infty(\Gamma)} < R$. Following the technique in the proof of the previous lemma, for sufficiently small R, first we absorb the term $\|\delta y\|_{L^2(\Omega)} + \|\delta y\|_{L^2(\Gamma)}$ in the left hand side

$$\|\delta y\|_{L^2(\Omega)} + \|\delta y\|_{L^2(\Gamma)} \leqslant c'(R)(\|\hat{y}\|_{L^2(\Omega)} + \|\hat{y}\|_{L^2(\Gamma)}).$$

Now the embedding and trace theorems imply

$$\|\delta y\|_{L^2(\Omega)} \leqslant c''(R) \|\hat{y}\|_{L^2(\Omega)},$$

where $c'(R)$ and $c''(R)$ have the same property as $c(R)$. Again, Lemma 8.2.3 implies that there exists $\alpha_0 > 0$ and $R > 0$ such that

$$\mathcal{L}_{xx}(w^*)(x, x) \geqslant \alpha_0 \|x\|^2_{L^2(\Omega) \times L^2(\Gamma)},$$

provided that $\|\tilde{y} - y^*\|_{L^\infty(\Omega)} < R$.

Owing to Lipschitz property (4.2.2), we further conclude that

$$\mathcal{L}_{xx}(\tilde{w})(x, x) = \mathcal{L}_{xx}(w^*)(x, x) + \big[\mathcal{L}_{xx}(\tilde{w}) - \mathcal{L}_{xx}(w^*)\big](x, x)$$
$$\geqslant \alpha_0 \|x\|^2_{L^2(\Omega) \times L^2(\Gamma)} - L \|\tilde{w} - w^*\|_{W^\infty} \|x\|^2_{L^2(\Omega) \times L^2(\Gamma)}$$
$$\geqslant (\alpha_0 - L R) \|x\|^2_{L^2(\Omega) \times L^2(\Gamma)} =: \alpha'' \|x\|^2_{L^2(\Omega) \times L^2(\Gamma)},$$

given that $w \in B_R^{W^\infty}(w^*)$. For sufficiently small R, we obtain $\alpha'' > 0$, which completes the proof. $\square$

Lemmas 4.2.3–4.2.4 argue the second order growth condition (4.0.3), which implies that **SSC** is sufficient for local optimality of x^*.

Theorem 4.2.5. *Under Assumptions* **(A1)-(A7)**, *there exist* $\beta > 0$ *and* $\varepsilon > 0$ *such that*

$$J(x) \geqslant J(x^*) + \beta \|x - x^*\|^2_{L^2(\Omega) \times L^2(\Gamma)} \qquad (4.2.8)$$

holds for all admissible $x \in X$ *with* $\|x - x^*\|_{L^\infty(\Omega) \times L^\infty(\Gamma)} \leqslant \varepsilon$. *In particular,* x^* *is a strict local optimal solution in the sense of* $L^\infty(\Omega) \times L^\infty(\Gamma)$.

Proof. Under Assumptions **(A2)-(A3)**, due to Lemma 4.2.3 there exist $\varepsilon > 0$ and $\alpha' > 0$ such that

$$\mathcal{L}_{xx}(x^*, \lambda^*)(x - x^*, x - x^*) \geqslant \alpha' \|x - x^*\|^2_{L^2(\Omega) \times L^2(\Gamma)}$$

for all $x - x^* = (y - y^*, u - u^*) \in X$ which satisfy

$$\mathcal{A}(y - y^*) + d(y) - d(y^*) = 0 \quad \text{in } \Omega$$
$$\partial_n(y - y^*) + b(y) - b(y^*) = (u - u^*) \quad \text{on } \Gamma.$$

provided that $\|x - x^*\|_{L^\infty(\Omega) \times L^\infty(\Gamma)} \leqslant \varepsilon$. This implies

$$\mathcal{L}(x, \lambda^*) \geqslant \mathcal{L}(x^*, \lambda^*) + \mathcal{L}_x(x^*, \lambda^*)(x - x^*) + \alpha' \|x - x^*\|^2_{L^2(\Omega) \times L^2(\Gamma)}$$
$$+ \left(\mathcal{L}_{xx}(x^* + \theta(x - x^*), \lambda^*) - \mathcal{L}_{xx}(x^*, \lambda^*)\right)(x - x^*, x - x^*)$$

for $\theta \in (0, 1)$. We remark $\mathcal{L}(x^*, \lambda^*) = J(x^*)$, $\mathcal{L}(x, \lambda^*) = J(x)$ and $\mathcal{L}_x(x^*, \lambda^*) = 0$. Moreover, the Hessian of the Lagrange functional satisfies the local Lipschitz condition (4.2.2). Thus, we obtain

$$J(x) \geqslant J(x^*) + \beta \|x - x^*\|^2_{L^2(\Omega) \times L^2(\Gamma)},$$

where

$$\beta := \alpha' - c \|x - x^*\|_{L^\infty(\Omega) \times L^\infty(\Gamma)} \geqslant \alpha' - c\varepsilon > 0$$

when ε is taken sufficiently small. $\square$

4.3 Distributed control problem

The study of *distributed control problems* (**P'**) follows the technique described in the previous sections. One only needs to take attention that the control u is not given on the boundary Γ, but distributed over whole domain Ω^1, i.e., $u \in L^\infty(\Omega)$, and pointwise inequality constraints are given a.e. in Ω. We recall

$$\text{Minimize} \quad J(y, u) := \int_\Omega \psi(\xi, y(\xi), u(\xi)) \, d\xi \qquad (\mathbf{P'})$$

subject to $u \in L^\infty(\Omega)$, and nonlinear elliptic state equation

$$\begin{aligned} \mathcal{A}y + d(y) &= u \quad \text{in } \Omega, \\ y &= 0 \quad \text{on } \Gamma \end{aligned} \qquad (4.3.1)$$

as well as pointwise nonlinear mixed constraints

$$g_i(y, u) \leqslant 0 \text{ a.e. in } \Omega, \quad i = 1, \ldots, s. \qquad (4.3.2)$$

---
[1]Assumptions for the distributed control problem (**P'**) are denoted by (**A1'**)–(**A7'**) and placed in Appendix 9.

4 Optimal control problem

Thanks to Lemma 8.8.6 for every $u \in L^\infty(\Omega)$ the state equation (4.3.1) possesses a unique solution $y \in W_0^{1,p}(\Omega)$, $p \in [1,\infty)$. (Note $L^\infty(\Omega) \hookrightarrow L^N(\Omega)$ owing to Theorem 8.5.4.) Moreover, due to Lemma 8.9.3 the unique weak solution y of (4.3.1) belongs to $H_0^1(\Omega) \cap H^2(\Omega)$. On one side, analogous to analysis for the boundary control problem (**P**) we can fix $\bar{p} > N$ and use the regularity of $W^{1,\bar{p}}(\Omega)$ for the state variable y, which should be sufficient for the distributed control problem too. On the other side, the H^2-regularity allows to simplify the verification of the stability estimate and the local quadratic convergence rate for the SQP method applied to (**P'**), for details see [5, 21].

From now on, for (**P'**) let $y \in Y$, where Y is either $W_0^{1,\bar{p}}(\Omega)$ (the proofs follow analogously to boundary control problem) or $H^2(\Omega) \cap H_0^1(\Omega)$ (for the proof we refer to [21]).

Similar to analysis for boundary control problem in Section 4.1.2, one shows the existence of regular multipliers $p \in Y$ and $\mu_i \in L^\infty(\Omega)$, $i = 1, \ldots, s$, cf. [40]. Therefore, similar to (4.1.5) we define the Lagrange functional $\mathcal{L} : Y \times L^\infty(\Omega) \times Y \times [L^\infty(\Omega)]^s \to \mathbb{R}$ by

$$\mathcal{L}(y, u, p, \mu_1, \ldots, \mu_s) = J(y, u) + a[y, p] + (d(y) - u, p)_\Omega + \sum_{i=1}^s (\mu_i, g_i(y, u))_\Omega.$$

Then first-order optimality conditions can be written like (4.1.6a)–(4.1.6d), where the test function v has to be chosen from $H_0^1(\Omega)$ and the gradient equation (4.1.6b) as well as complementary conditions (4.1.6d) have to be considered a.e. in Ω. Again, as we see in Example 4.3.1 below, if μ_i, $i = 1, \ldots, s$ are simultaneously nonzero on a set of positive measure, the Lagrange multipliers $(p, \mu_1, \ldots, \mu_s)$ fail to be unique.

Example 4.3.1.

$$\text{Minimize } \frac{1}{2}\|y\|_{L^2(\Omega)}^2 + \frac{\gamma}{2}\|u - u_d\|_{L^2(\Omega)}^2$$

$$\text{subject to } \begin{cases} -\Delta y = u & \text{in } \Omega, \quad u \leq 0 \quad \text{in } \Omega \\ y = 0 & \text{on } \Gamma, \quad \varepsilon u + y \leq 0 \quad \text{in } \Omega. \end{cases}$$

Suppose that $u_d := \gamma^{-1}(\varepsilon + S1)$, where $S1$ denotes the solution of the Laplace equation with right hand side equal 1. Due to the maximum principle (see Corollary 8.7.4 and Lemma 8.8.2), $-u_d \leq -\gamma^{-1}\varepsilon$ holds a.e. on Ω. Apparently, $y = u = 0$ is the unique solution of this problem. Any tuple (p, μ_1, μ_2) satisfying

4.3 Distributed control problem

(4.1.6a), (4.1.6b) and (4.1.6d), i.e.,

$$-\Delta p = -\mu_2 \quad in \ \Omega \qquad \mu_1 \geqslant 0, \quad \mu_2 \geqslant 0 \quad a.e. \ in \ \Omega$$
$$p = 0 \quad on \ \Gamma \qquad -\gamma u_d - p + \mu_1 + \varepsilon\mu_2 = 0 \quad a.e. \ in \ \Omega$$

is a set of Lagrange multipliers for the problem. It is easy to check that $(p, \mu_1, \mu_2) = (-S1, 0, 1)$ and $(p, \mu_1, \mu_2) = (0, \varepsilon + S1, 0)$ both satisfy this system, and so does any convex combination.

One can avoid such a case by introducing the security sets in Ω (according to Definition 4.1.5 and modified with respect to problem setting (**P'**)) and assuming (**A6'**), see Appendix 9.

Theorem 4.3.2. *If Assumption* (**A1'**)–(**A6'**) *are satisfied, the Lagrange multipliers and the adjoint state are unique.*

The second-order sufficient condition for (**P'**) presented by Assumption (**A7'**) guarantees a critical point $x^* \in Y \times L^\infty(\Omega)$ to be a strict local optimizer for (**P'**).

Theorem 4.3.3. *Under Assumptions* (**A1'**)–(**A7'**), *there exist* $\beta > 0$ *and* $\varepsilon > 0$ *such that*

$$J(x) \geqslant J(x^*) + \beta \|x - x^*\|^2_{[L^2(\Omega)]^2}$$

holds for all admissible $x \in Y \times L^\infty(\Omega)$ *with* $\|x - x^*\|_{[L^\infty(\Omega)]^2} \leqslant \varepsilon$. *In particular, x^* is a strict local optimal solution in the sense of* $[L^\infty(\Omega)]^2$.

4 Optimal control problem

5 Stability of linear-quadratic problems

In Chapter 3 we have seen the feature of the SQP method is to solve the quadratic subproblem in each iteration. The SQP method provides the local quadratic convergence of iterates if one affirms the stability of the solutions of the subproblems and stability of the active sets. The aim of this chapter is to confirm Lipschitz stability of solutions of linear-quadratic subproblems associated with (**P**). We will also show the stability of Lagrange multipliers associated to this solution, which provides the stability of active sets, see Theorem 5.2.1 and Corollary 5.2.2 below. We recall

$$W := W^{1,\bar{p}}(\Omega) \times L^\infty(\Gamma) \times W^{1,\bar{p}}(\Omega) \times [L^\infty(\Gamma)]^s \text{ with } \bar{p} \in [N, \infty)$$

and for the further discussion, we define the space

$$Z := W^{1,\bar{p}'}(\Omega)^* \times L^\infty(\Gamma) \times W^{1,\bar{p}'}(\Omega)^* \times [L^\infty(\Gamma)]^s,$$

where $\bar{p}'$ is conjugate to $\bar{p}$, i.e., $1/\bar{p} + 1/\bar{p}' = 1$.

Let $w^* = (y^*, u^*, p^*, \mu_1^*, \ldots, \mu_s^*) \in W$ be an element, which satisfies the first order necessary optimality conditions (4.1.6a)–(4.1.6d) as well as Assumption (**A7**), and x^* abbreviates (y^*, u^*). We consider the following linear-quadratic problem which depends on perturbation $\delta \in Z$

$$\text{Minimize} \quad J_x(x^*)(x - x^*) + \frac{1}{2}\mathcal{L}_{xx}(w^*)(x - x^*, x - x^*) \quad \text{(\textbf{LQP}}(\delta)\text{)}$$
$$- \langle \delta_1, y - y^* \rangle_{W^{1,\bar{p}'}(\Omega)^*, W^{1,\bar{p}}(\Omega)} - (\delta_2, u - u^*)_\Gamma$$

subject to $u \in L^\infty(\Gamma)$, the linearized state equation

$$\begin{aligned}\mathcal{A}y + d(y^*) + d_y(y^*)(y - y^*) &= f + \delta_{3,\Omega} & \text{in } \Omega, \\ \partial_n y + b(y^*) + b_y(y^*)(y - y^*) &= u + \delta_{3,\Gamma} & \text{on } \Gamma\end{aligned} \quad (5.0.1)$$

as well as the linearized inequality constraints

$$g_i(x^*) + g_{i,x}(x^*)(x - x^*) \leqslant \delta_{i+3} \qquad \text{a.e. on } \Gamma, \qquad (5.0.2)$$
$$i = 1, \ldots, s.$$

5 Stability of linear-quadratic problems

The necessary optimality conditions for **LQP**(δ) can be obtained by linearizing (4.1.6a)–(4.1.6d), which leads to

$$a[v,p] + (d_y(y^*)p, v)_\Omega + (d_{yy}(y^*)(y-y^*)p^*, v)_\Omega + (b_y(y^*)p, v)_\Gamma$$
$$+ (b_{yy}(y^*)(y-y^*)p^*, v)_\Gamma + (\phi_y(y^*), v)_\Omega + (\phi_{yy}(y^*)(y-y^*), v)_\Omega$$
$$+ (\psi_y(y^*, u^*), v)_\Gamma + (\psi_{yy}(y^*, u^*)(y-y^*), v)_\Gamma + (\psi_{yu}(y^*, u^*)(u-u^*), v)_\Gamma$$
$$+ \sum_{i=1}^{s}(g_{i,y}(y^*, u^*)\mu_i, v)_\Gamma + \sum_{i=1}^{s}(g_{i,yy}(y^*, u^*)(y-y^*)\mu_i^*, v)_\Gamma$$
$$+ \sum_{i=1}^{s}(g_{i,yu}(y^*, u^*)(u-u^*)\mu_i^*, v)_\Gamma = \langle \delta_1, v \rangle \quad \text{for all } v \in W^{1,\bar{p}'}(\Omega)$$

(5.0.3a)

$$\psi_u(y^*, u^*) + \psi_{uy}(y^*, u^*)(y-y^*) + \psi_{uu}(y^*, u^*)(u-u^*) - p + \sum_{i=1}^{s} g_{i,u}(y^*, u^*)\mu_i$$
$$+ \sum_{i=1}^{s} g_{i,uy}(y^*, u^*)(y-y^*)\mu_i^* + \sum_{i=1}^{s} g_{i,uu}(y^*, u^*)(u-u^*)\mu_i^* = \delta_2, \quad \text{a.e. on } \Gamma$$

(5.0.3b)

$$a[y,v] + (d(y^*), v)_\Omega + (d_y(y^*)(y-y^*), v)_\Omega + (b(y^*), v)_\Gamma + (b_y(y^*)(y-y^*), v)_\Gamma$$
$$- (f, v)_\Omega - (u, v)_\Gamma = \langle \delta_3, v \rangle \quad \text{for all } v \in W^{1,\bar{p}'}(\Omega)$$

(5.0.3c)

$$0 \leqslant \mu_i \perp g_i(x^*) + g_{i,x}(x^*)(x-x^*) \leqslant \delta_{i+3}, \quad \text{a.e. on } \Gamma, \ i = 1, \ldots, s. \quad (5.0.3d)$$

Following the technique in Section 4.1.2 we obtain the existence of regular Lagrange multipliers associated to **LQP**(δ). Assumption **(A6)** is necessary for their uniqueness. In order to show stability of solutions of **LQP**(δ) we follow the technique first proposed in [6] for optimal control of ODEs. In the context of our problem setting, that is,

(1) We define an auxiliary problem denoted by **LQP**^{aux}(δ) and prove the uniqueness and Lipschitz stability of solution for this problem w.r.t. L^2 space setting

(2) We devise the projection formula which helps with the extension of the stability result to W and Z

(3) We show that solutions of **LQP**(δ) and **LQP**^{aux}(δ) coincide for small $\delta \in Z$.

Section 5.1 connects to first and second items; the last step is shown in Section 5.2.

5.1 Stability result for auxiliary problem

We start with definition of the auxiliary linear-quadratic problem $(\mathbf{LQP^{aux}}(\delta))$, in which the inequality constraints (5.0.2) are restricted on disjoint sets S_i

$$\text{Minimize} \quad J_x(x^*)(x-x^*) + \frac{1}{2}\mathcal{L}_{xx}(w^*)(x-x^*, x-x^*) \quad (\mathbf{LQP^{aux}}(\delta))$$
$$- \langle \delta_1, y - y^* \rangle_{W^{1,\bar{p}'}(\Omega)^*, W^{1,\bar{p}}(\Omega)} - (\delta_2, u - u^*)_\Gamma$$

subject to $u \in L^\infty(\Gamma)$, the linearized state equation

$$\begin{aligned}
\mathcal{A}y + d(y^*) + d_y(y^*)(y - y^*) &= f + \delta_{3,\Omega} \quad \text{in } \Omega, \\
\partial_n y + b(y^*) + b_y(y^*)(y - y^*) &= u + \delta_{3,\Gamma} \quad \text{on } \Gamma
\end{aligned} \quad (5.1.1)$$

as well as the linearized inequality constraints

$$g_i(x^*) + g_{i,x}(x^*)(x - x^*) \leqslant \delta_{i+3} \quad \text{a.e. on } S_i, \quad (5.1.2)$$
$$i = 1, \ldots, s.$$

First we proceed to show the unique global solvability of $(\mathbf{LQP^{aux}}(\delta))$ and the Lipschitz stability of its solution in the spaces

$$W_0 = H^1(\Omega) \times L^2(\Gamma) \times H^1(\Omega) \times [L^2(\Gamma)]^s,$$
$$Z_0 = H^1(\Omega)^* \times L^2(\Gamma) \times H^1(\Omega)^* \times [L^2(\Gamma)]^s.$$

Note, the embedding theorem implies $Z \hookrightarrow Z_0$ as well as $W \hookrightarrow W_0$. Moreover, while the set of feasible pairs for $(\mathbf{LQP}(\delta))$ could be empty in general, thanks to the separation assumption (**A6**) the corresponding auxiliary problem $(\mathbf{LQP^{aux}}(\delta))$ always possesses at least one point $x \in X$ satisfying (5.1.1) and (5.1.2).

Lemma 5.1.1. *Suppose that Assumption* (**A1**)-(**A7**) *hold. Then for any* $\delta \in Z_0$ *the set of feasible pairs*

$$M_\delta^{aux} = \{(y, u) \in H^1(\Omega) \times L^2(\Gamma) \text{ satisfying } (5.1.1) \text{ and } (5.1.2)\}$$

is nonempty.

Proof. We will show that there exists a pair $(y, u) \in H^1(\Omega) \times L^2(\Gamma)$ such that

$$g_i(x^*) + g_{i,x}(x^*)(x - x^*) = \delta_{i+3} \quad \text{a.e. on } S_i, \; i = 1, \ldots, s,$$

and (5.1.1) are satisfied, i.e., all inequality constraints hold with equality, which

5 Stability of linear-quadratic problems

implies $(y, u) \in M_\delta^{aux}$. We set

$$u := \begin{cases} g_{i,u}^{-1}(y^*, u^*)[\delta_{i+3} - g_i(y^*, u^*) - g_{i,y}(y^*, u^*)(y - y^*)] + u^*, & \text{on } S_i, \\ & i = 1, \ldots, s \\ 0, & \text{on } \Gamma \setminus \bigcup_{i=1}^{s} S_i. \end{cases} \quad (5.1.3)$$

Plugging this into the state equation (5.1.1) we find

$$\mathcal{A}y + d_y(y^*)\, y = \tilde{r}_1 \quad \text{in } \Omega, \qquad (5.1.4)$$

$$\partial_n y + [b_y(y^*) + \sum_{i=1}^{s} \chi_{S_i}\, g_{i,u}^{-1}(y^*, u^*)\, g_{i,y}(y^*, u^*)]\, y = \tilde{r}_2 \quad \text{on } \Gamma,$$

where $\tilde{r}_1 \in L^2(\Omega)$ and $\tilde{r}_2 \in L^2(\Gamma)$ are given by

$$\tilde{r}_1 = f + \delta_{3,\Omega} - d(y^*) + d_y(y^*)\, y^*$$
$$\tilde{r}_2 = u^* + \delta_{3,\Gamma} - b(y^*) + b_y(y^*)\, y^*$$
$$+ \sum_{i=1}^{s} \chi_{S_i}\, g_{i,u}^{-1}(y^*, u^*)[g_{i,y}(y^*, u^*)\, y^* - g_i(y^*, u^*) + \delta_{i+3}]$$

It remains to show that there exists a (unique) weak solution $y \in H^1(\Omega)$ of (5.1.4) satisfying

$$a[y, v] + (d_y(y^*)\, y, v)_\Omega + (b_y(y^*)\, y, v)_\Gamma + \sum_{i=1}^{s} (\chi_{S_i}\, g_{i,u}^{-1}(y^*, u^*)\, g_{i,y}(y^*, u^*)\, y, v)_\Gamma$$
$$= (\tilde{r}_1, v)_\Omega + (\tilde{r}_2, v)_\Gamma \quad (5.1.5)$$

for all $v \in H^1(\Omega)$ and arbitrary $\tilde{r}_1 \in L^2(\Omega)$ and $\tilde{r}_2 \in L^2(\Gamma)$. By (5.1.3), this implies $u \in L^2(\Omega)$.

We consider (5.1.5) for $\tilde{r}_1 = 0$ and $\tilde{r}_2 = 0$. Let $y \in H^1(\Omega)$ be an arbitrary solution of this homogeneous equation (at least $y \equiv 0$ is valid). Let $r_1 \in L^2(\Omega)$ and $r_2 \in L^2(\Gamma)$ be arbitrary. By Assumption (**A6**), we obtain the existence of $p = p(r_1, r_2) \in H^1(\Omega)$ such that

$$a[v, p] + (d_y(y^*)\, p, v)_\Omega + (b_y(y^*)\, p, v)_\Gamma + \sum_{i=1}^{s} (\chi_{S_i}\, g_{i,u}^{-1}(y^*, u^*)\, g_{i,y}(y^*, u^*)\, p, v)_\Gamma$$
$$= (r_1, v)_\Omega + (r_2, v)_\Gamma \quad (5.1.6)$$

for all $v \in H^1(\Omega)$. We set $v := y$ in (5.1.6) and obtain by (5.1.5)

$$0 = (r_1, y)_\Omega + (r_2, y)_\Gamma.$$

5.1 Stability result for auxiliary problem

Since r_1 and r_2 were arbitrary, this implies $y = 0$, and the homogeneous version of (5.1.5) is uniquely solvable. Thus due to Fredholm theorem [19, Theorem D.5] the state equation (5.1.5) has a unique solution for arbitrary $\bar{r}_1 \in L^2(\Omega)$ and $\bar{r}_2 \in L^2(\Gamma)$. $\square$

The necessary optimality conditions for (**LQP^{aux}**(δ)) are (5.0.3a)–(5.0.3c) and

$$0 \leqslant \mu_i \perp g_i(x^*) + g_{i,x}(x^*)(x - x^*) \leqslant \delta_{i+3}, \quad \text{a.e. on } S_i, \ i = 1,\ldots,s. \quad (5.1.7)$$

Lemma 5.1.2. *Suppose that Assumption* (**A1**)–(**A7**) *hold. For every* $\delta \in Z_0$, *the auxiliary problem* (**LQP^{aux}**(δ)) *possesses a unique global solution* $(y_\delta^{aux}, u_\delta^{aux}) \in H^1(\Omega) \times L^2(\Gamma)$. *Moreover, the associated adjoint state* $p_\delta^{aux} \in H^1(\Omega)$ *and Lagrange multipliers* $\mu_{i,\delta}^{aux} \in L^2(\Gamma)$ *are unique.*

Proof. By the previous lemma, the set of feasible pairs M_δ^{aux} is nonempty, and it is also closed and convex. Recall that x^* satisfies the state equation (4.0.1). We denote $\delta u := u - u^*$, $\delta y \in H^1(\Omega)$ is the corresponding solution of the linearized PDE (4.2.4). Then any $x \in M_\delta^{aux}$ can be presented as $x = x^* + \delta x + e_x$, where the remaining term $e_x := (e_y, 0)$ satisfies

$$\begin{aligned}
\mathcal{A}\, e_y + d_y(y^*)\, e_y &= \delta_{3,\Omega} \quad \text{in } \Omega, \\
\partial_n e_y + b_y(y^*)\, e_y &= \delta_{3,\Gamma} \quad \text{on } \Gamma.
\end{aligned} \quad (5.1.8)$$

Thanks to Assumption (**A7**) there exists a constant $\alpha > 0$ such that

$$\begin{aligned}
\mathcal{L}_{xx}(w^*)(x - x^*, x - x^*) &= \mathcal{L}_{xx}(w^*)(\delta x + e_x, \delta x + e_x) \\
&= \mathcal{L}_{xx}(w^*)(\delta x, \delta x) + 2\mathcal{L}_{xx}(w^*)(\delta x, e_x) + \mathcal{L}_{xx}(w^*)(e_x, e_x) \\
&\overset{(\mathbf{A7})}{\geqslant} \alpha \, \|\delta x\|^2_{L^2(\Omega) \times L^2(\Gamma)} + 2\mathcal{L}_{xx}(w^*)(\delta x, e_x) + \mathcal{L}_{xx}(w^*)(e_x, e_x) \\
&\overset{\delta x = x - x^* - e_x}{\geqslant} \alpha \, \|x\|^2_{L^2(\Omega) \times L^2(\Gamma)} + \text{terms affine w.r.t. } x
\end{aligned}$$

for all $x \in M_\delta^{aux}$. Hence, the objective of (**LQP^{aux}**(δ)) is strictly convex and weakly lower semicontinuous, and it is also radially unbounded. This shows that (**LQP^{aux}**(δ)) has a unique solution $(y_\delta^{aux}, u_\delta^{aux}) \in H^1(\Omega) \times L^2(\Gamma)$.

Following the techniques in Section 4.1.2, and thanks to the linearized Slater condition (**A4**), one can show the existence of Lagrange multipliers $\mu_{1,\delta}^{aux}, \ldots, \mu_{s,\delta}^{aux}$ first in $L^\infty(\Gamma)^*$, then in $L^2(\Gamma)$, and existence of an adjoint state $p_\delta^{aux} \in H^1(\Omega)$, such that the optimality conditions (5.0.3a)-(5.0.3c) and (5.1.7) are satisfied. $\square$

Theorem 5.1.4 below provides the Lipschitz stability result for (**LQP^{aux}**(δ)) with respect to L^2 space setting. For its proof, we need the following auxiliary

5 Stability of linear-quadratic problems

result. Note the Lagrange multipliers μ_i^{aux} are well-defined on S_i from (5.1.7) and extended by zero on $\Gamma \setminus S_i$. ($i = 1, \ldots, s$)

Lemma 5.1.3. *The Lagrange multipliers associated to the solution x_δ and $x_{\delta'}$ of* (**LQP**$^{aux}(\delta)$) *and* (**LQP**$^{aux}(\delta')$), *respectively, satisfy the inequality*

$$\sum_{i=1}^{s} \left(\mu_{i,\delta}^{aux} - \mu_{i,\delta'}^{aux}, g_{i,x}(x^*)(x_\delta^{aux} - x_{\delta'}^{aux})\right)_\Gamma \geq \sum_{i=1}^{s} (\mu_{i,\delta}^{aux} - \mu_{i,\delta'}^{aux}, \delta_{i+3} - \delta'_{i+3})_\Gamma.$$

Proof. Using the complementarity conditions (5.0.3d), we infer

$$-\mu_{1,\delta}^{aux}(g_1(x^*) + g_{1,x}(x^*)(x_\delta^{aux} - x^*) - \delta_4) = 0 \quad \text{a.e. on } S_1,$$
$$\mu_{1,\delta'}^{aux}(g_1(x^*) + g_{1,x}(x^*)(x_\delta^{aux} - x^*) - \delta_4) \leq 0 \quad \text{a.e. on } \Gamma$$

and

$$-\mu_{1,\delta'}^{aux}(g_1(x^*) + g_{1,x}(x^*)(x_{\delta'}^{aux} - x^*) - \delta'_4) = 0 \quad \text{a.e. on } S_1,$$
$$\mu_{1,\delta}^{aux}(g_1(x^*) + g_{1,x}(x^*)(x_{\delta'}^{aux} - x^*) - \delta'_4) \leq 0 \quad \text{a.e. on } \Gamma.$$

Therefore

$$\left(\mu_{1,\delta}^{aux} - \mu_{1,\delta'}^{aux}, g_{1,x}(x^*)(x_\delta^{aux} - x_{\delta'}^{aux})\right)_\Gamma \geq \left(\mu_{1,\delta}^{aux} - \mu_{1,\delta'}^{aux}, \delta_4 - \delta'_4\right)_\Gamma$$

follows. Similarly, one obtains the remainder parts. $\square$

Theorem 5.1.4. *Suppose that Assumptions* (**A1**)-(**A7**) *hold. Then there exists a constant $L_0 > 0$ such that*

$$\|w_\delta^{aux} - w_{\delta'}^{aux}\|_{W_0} \leq L_0 \, \|\delta - \delta'\|_{Z_0} \qquad (5.1.9)$$

holds for all $\delta, \delta' \in Z_0$.

Proof. Let $\delta w = (\delta y, \delta u, \delta p, \delta \mu_1, \ldots, \delta \mu_s)$ denote the difference $w_\delta^{aux} - w_{\delta'}^{aux}$.

40

5.1 Stability result for auxiliary problem

The quantities δp, δu and δy satisfy equations

$$a[v, \delta p] + (d_y(y^*)\delta p, v)_\Omega + (b_y(y^*)\delta p, v)_\Gamma + (d_{yy}(y^*)p^*\delta y, v)_\Omega + (b_{yy}(y^*)p^*\delta y, v)_\Gamma$$
$$+ (\phi_{yy}(y^*)\delta y, v)_\Omega + (\psi_{yy}(y^*, u^*)\delta y, v)_\Gamma + (\psi_{yu}(y^*, u^*)\delta u, v)_\Gamma$$
$$+ \sum_{i=1}^{s}(g_{i,yy}(y^*, u^*)\mu_i^*\delta y, v)_\Gamma + \sum_{i=1}^{s}(g_{i,yu}(y^*, u^*)\mu_i^*\delta u, v)_\Gamma$$
$$+ \sum_{i=1}^{s}(g_{i,y}(y^*, u^*)\delta\mu_i, v)_\Gamma = \langle \delta_1 - \delta_1', v \rangle_{H^1(\Omega)^*, H^1(\Omega)} \qquad v \in H^1(\Omega)$$

$$\left[\psi_{uy}(y^*, u^*) + \sum_{i=1}^{s} g_{i,uy}(y^*, u^*)\mu_i^*\right]\delta y + \left[\psi_{uu}(y^*, u^*) + \sum_{i=1}^{s} g_{i,uu}(y^*, u^*)\mu_i^*\right]\delta u$$
$$- \delta p + \sum_{i=1}^{s} g_{i,u}(y^*, u^*)\delta\mu_i = \delta_2 - \delta_2' \qquad \text{a.e. on } \Gamma$$

$$a[\delta y, v] + (d_y(y^*)\delta y, v)_\Omega + (b_y(y^*)\delta y, v)_\Gamma - (\delta u, v)_\Gamma = \langle \delta_3 - \delta_3', v \rangle_{H^1(\Omega)^*, H^1(\Omega)}$$
$$v \in H^1(\Omega)$$

Testing the latter equations with δy, δu and $-\delta p$, respectively, and adding up, we obtain

$$\mathcal{L}_{xx}(w^*)(\delta x, \delta x) + \sum_{i=1}^{s}(g_{i,x}(x^*)\delta\mu_i, \delta x)_\Gamma = \langle \delta_1 - \delta_1', \delta y \rangle_{H^1(\Omega)^*, H^1(\Omega)} + (\delta_2 - \delta_2', \delta u)_\Gamma$$
$$- \langle \delta_3 - \delta_3', \delta p \rangle_{H^1(\Omega)^*, H^1(\Omega)}$$

Due to Lemma 5.1.3 $\sum_{i=1}^{s}(g_{i,x}(x^*)\delta\mu_i, \delta x)_\Gamma \geqslant \sum_{i=1}^{s}(\delta\mu_i, \delta_{i+3} - \delta_{i+3}')_\Gamma$ holds. Using the Cauchy-Schwarz inequality, we estimate

$$\mathcal{L}_{xx}(w^*)(\delta x, \delta x) \leqslant \|\delta_1 - \delta_1'\|_{H^1(\Omega)^*} \|\delta y\|_{H^1(\Omega)} + \|\delta_2 - \delta_2'\|_{L^2(\Gamma)} \|\delta u\|_{L^2(\Gamma)}$$
$$+ \|\delta_3 - \delta_3'\|_{H^1(\Omega)^*} \|\delta p\|_{H^1(\Omega)} + \sum_{i=1}^{s} \|\delta_{i+3} - \delta_{i+3}'\|_{L^2(\Gamma)} \|\delta\mu_i\|_{L^2(\Gamma)}$$
$$(5.1.10)$$

We follow the technique in Lemma 5.1.2 to appraise the value of $\mathcal{L}_{xx}(w^*)(\delta x, \delta x)$. We split $\delta x := \hat{x} + e_x$, where $\hat{x}$ satisfies (4.2.4) and e_x satisfies (5.1.8), and find

$$\mathcal{L}_{xx}(w^*)(\delta x, \delta x) = \mathcal{L}_{xx}(w^*)(\hat{x} + e_x, \hat{x} + e_x)$$
$$= \mathcal{L}_{xx}(w^*)(\hat{x}, \hat{x}) + 2\mathcal{L}_{xx}(w^*)(\hat{x}, e_x) + \mathcal{L}_{xx}(w^*)(e_x, e_x)$$
$$\overset{(A7)}{\geqslant} \alpha \|\hat{x}\|^2_{L^2(\Omega) \times L^2(\Gamma)} - 2\|\mathcal{L}_{xx}(w^*)\| \|\hat{x}\|_{L^2(\Omega) \times L^2(\Gamma)} \|e_x\|_{L^2(\Omega) \times L^2(\Gamma)}$$
$$- \|\mathcal{L}_{xx}(w^*)\| \|e_x\|^2_{L^2(\Omega) \times L^2(\Gamma)}$$

5 Stability of linear-quadratic problems

Using that $\hat{x} = \delta x + e_x$ and triangle's inequality the first summand values at

$$\alpha \|\delta x\|^2_{L^2(\Omega) \times L^2(\Gamma)} - 2\alpha \|\delta x\|_{L^2(\Omega) \times L^2(\Gamma)} \|e_x\|_{L^2(\Omega) \times L^2(\Gamma)} + \alpha \|e_x\|^2_{L^2(\Omega) \times L^2(\Gamma)}.$$

Applying Young's inequality (8.0.3) with $\gamma_1 > 0$ and $\gamma_2 > 0$ we estimate

$$\alpha \|\delta x\|_{L^2(\Omega) \times L^2(\Gamma)} \|e_x\|_{L^2(\Omega) \times L^2(\Gamma)}$$
$$\leqslant \alpha \gamma_1 \|\delta x\|^2_{L^2(\Omega) \times L^2(\Gamma)} + \alpha C(\gamma_1) \|e_x\|^2_{L^2(\Omega) \times L^2(\Gamma)}$$

and

$$\|\mathcal{L}_{xx}(w^*)\| \|\delta x\|_{L^2(\Omega) \times L^2(\Gamma)} \|e_x\|_{L^2(\Omega) \times L^2(\Gamma)}$$
$$\leqslant \gamma_2 \|\mathcal{L}_{xx}(w^*)\| \|\delta x\|^2_{L^2(\Omega) \times L^2(\Gamma)} + C(\gamma_2) \|\mathcal{L}_{xx}(w^*)\| \|e_x\|^2_{L^2(\Omega) \times L^2(\Gamma)}.$$

Summarizing all intermediate steps we obtain

$$\mathcal{L}_{xx}(w^*)(\delta x, \delta x) \geqslant \alpha''' \|\delta x\|^2_{L^2(\Omega) \times L^2(\Gamma)} + c_0 \|e_x\|^2_{L^2(\Omega) \times L^2(\Gamma)},$$

where $\alpha''' := \alpha - 2\alpha\gamma_1 - 2\gamma_2 \|\mathcal{L}_{xx}(w^*)\|$ and $c_0 := \alpha - 2\alpha C(\gamma_1) - 2C(\gamma_2) \|\mathcal{L}_{xx}(w^*)\|$. Note γ_1 and γ_2 have to be chosen such that $\alpha''' > 0$.

Since $e_x = (e_y, 0)$ satisfies the linaer PDE (5.1.8) then the estimate

$$\|e_y\|_{L^2(\Omega)} \leqslant C_2 \|e_y\|_{H^1(\Omega)} \leqslant C_2 \hat{c}_\Omega \|\delta_3 - \delta_3'\|_{H^1(\Omega)^*}$$

holds and (5.1.10) provides

$$\alpha''' \|\delta x\|^2_{L^2(\Omega) \times L^2(\Gamma)} \leqslant \|\delta_1 - \delta_1'\|_{H^1(\Omega)^*} \|\delta y\|_{H^1(\Omega)} + \|\delta_2 - \delta_2'\|_{L^2(\Gamma)} \|\delta u\|_{L^2(\Gamma)}$$
$$+ \|\delta_3 - \delta_3'\|_{H^1(\Omega)^*} \|\delta p\|_{H^1(\Omega)} + \sum_{i=1}^{s} \|\delta_{i+3} - \delta'_{i+3}\|_{L^2(\Gamma)} \|\delta \mu_i\|_{L^2(\Gamma)}$$
$$+ C_2 \hat{c}_\Omega c_0 (\|\delta_3 - \delta_3\|_{H^1(\Omega)^*}) \tag{5.1.11}$$

From the gradient equation (5.0.3b) we have

$$\sum_{i=1}^{s} g_{i,u}(y^*, u^*) \delta \mu_i = \delta p + \delta_2 - \delta_2' - [\psi_{uu}(y^*, u^*) + \sum_{i=1}^{s} g_{i,uu}(y^*, u^*) \mu_i^*] \delta u$$
$$- [\psi_{uy}(y^*, u^*) + \sum_{i=1}^{s} g_{i,uy}(y^*, u^*) \mu_i^*] \delta y, \quad \text{a.e. on } \Gamma$$

Since $g_{i,u}(x^*) \geqslant m$, we estimate

$$\|\delta \mu_i\|_{L^2(\Gamma)} \leqslant c \big(\|\delta y\|_{L^2(\Gamma)} + \|\delta u\|_{L^2(\Gamma)} + \|\delta p\|_{L^2(\Gamma)} + \|\delta_2 - \delta_2'\|_{L^2(\Gamma)} \big).$$

Proceeding as in Theorem 4.1.6 and using a standard a priori estimate for the adjoint equation in **(A6)**, we get

$$\|\delta p\|_{H^1(\Omega)} \leqslant \tilde{c}\big(\|\delta y\|_{L^2(\Omega)}+\|\delta y\|_{L^2(\Gamma)}+\|\delta u\|_{L^2(\Gamma)}+\|\delta_1-\delta_1'\|_{H^1(\Omega)^*}+\|\delta_2-\delta_2'\|_{L^2(\Gamma)}\big).$$

Plugging these estimates into (5.1.11), we find

$$\alpha'''\,\|\delta x\|_{L^2(\Omega)\times L^2(\Gamma)}^2 \leqslant \big(\|\delta_1-\delta_1'\|_{H^1(\Omega)^*} + \tilde{c}\,\|\delta_3-\delta_3'\|_{H^1(\Omega)^*}\big)\,\|\delta y\|_{H^1(\Omega)}$$

$$+ \big(\tilde{c}\,\|\delta_3-\delta_3'\|_{H^1(\Omega)^*} + (c+c\tilde{c})\sum_{i=1}^{s}\|\delta_{i+3}-\delta_{i+3}'\|_{L^2(\Gamma)}\big)\,\|\delta y\|_{L^2(\Gamma)}$$

$$+ \big(\|\delta_2-\delta_2'\|_{L^2(\Gamma)} + \tilde{c}\,\|\delta_3-\delta_3'\|_{H^1(\Omega)^*} + (c+c\tilde{c})\sum_{i=1}^{s}\|\delta_{i+3}-\delta_{i+3}'\|_{L^2(\Gamma)}\big)\,\|\delta u\|_{L^2(\Gamma)}$$

$$+ c_1\,\|\delta-\delta'\|_{Z_0}^2 \qquad (c_1 \text{ is a positive constant})$$

The trace theorem 8.5.5 implies $H^1(\Omega) \hookrightarrow L^2(\Gamma)$, and $\|\delta y\|_{L^2(\Gamma)} \leqslant \hat{c}_\tau\,\|\delta y\|_{H^1(\Omega)}$ holds. Using $\|\delta y\|_{H^1(\Omega)} \leqslant \hat{c}_\Omega(\|\delta u\|_{L^2(\Gamma)} + \|\delta_3-\delta_3'\|_{H^1(\Omega)^*})$, we obtain

$$\alpha'''\,\|\delta x\|_{L^2(\Omega)\times L^2(\Gamma)}^2 \leqslant \hat{c}\,\|\delta-\delta'\|_{Z_0}\,\|\delta u\|_{L^2(\Gamma)} + c_1\,\|\delta-\delta'\|_{Z_0}^2.$$

Finally, Young's inequality (8.0.3) yields the estimate

$$\|\delta y\|_{L^2(\Omega)}^2 + (\alpha''' - \frac{\hat{c}}{4\gamma})\,\|\delta u\|_{L^2(\Gamma)}^2 \leqslant \hat{c}\gamma\,\|\delta-\delta'\|_{Z_0}^2 + c_1\,\|\delta-\delta'\|_{Z_0}^2,$$

where $\alpha''' - \frac{\hat{c}}{4\gamma} > 0$ by appropriate choice of the constant $\gamma > 0$. Thus, we have

$$\|\delta y\|_{L^2(\Omega)}^2 + \|\delta u\|_{L^2(\Gamma)}^2 \leqslant L\,\|\delta-\delta'\|_{Z_0}^2$$

with $L > 0$. The estimates for δy, δp and $\delta \mu_i$ above conclude the proof. □

Theorem 5.1.4 describes the stability behavior for the auxiliary problem (**LQP**$^{\text{aux}}(\delta)$). However, the results are not strong enough to apply them for the original problem (**LQP**(δ)), see Proposition 5.1.6 below. First we define the active sets for (**P**).

Definition 5.1.5. The <u>active set</u> $\mathcal{A}_i$ at point $(y^*, u^*) \in X$ is defined by

$$\mathcal{A}_i = \{\xi \in \Gamma : \ g_i(\xi, y^*(\xi), u^*(\xi)) = 0\}, \ i = 1, \ldots, s.$$

Proposition 5.1.6. *Suppose that Assumptions* (**A1**)–(**A3**) *hold and that x^* satisfies the separation assumption* (**A6**). *Moreover, we assume that the active set $\mathcal{A}_s$ contains an open ball B such that $\mu_s^* \geqslant M > 0$ holds on B. Then for every $R > 0$ there exists $\delta \in Z_0$ with $\|\delta\|_{Z_0} < R$ such that the dual variables for*

5 Stability of linear-quadratic problems

(**LQP**(δ)) are not unique. Consequently, the dual variables cannot be Lipschitz stable with respect to the perturbations in Z_0.

Proof. Recall that $w^* = (y^*, u^*, p^*, \mu_1^*, \ldots, \mu_s^*)$ is a solution of the optimality system (5.0.3a)–(5.0.3d) for $\delta \equiv 0$. Due to the separation assumption (**A6**), w^* is also a solution of the optimality system for the auxiliary problem (**LQP**^{aux}(0)). Since the solution of the optimality system for (**LQP**^{aux}(0)) is unique, see Lemma 5.1.2, this uniqueness must hold for (**LQP**(0)) as well. In particular, $(y^*, u^*, p^*, \mu_1^*, \ldots, \mu_s^*) = (y^{*\text{aux}}, u^{*\text{aux}}, p^{*\text{aux}}, \mu_1^{*\text{aux}}, \ldots, \mu_s^{*\text{aux}})$.

Let us denote by B the open ball centered at $\xi \in \Gamma$ contained in A_s such that $\mu_s^* \geqslant M > 0$ holds on B. Let $r > 0$ such that $B_r(\xi) \subset B$ and $\max_i \|g_i(x^*)\|_{L^\infty(\Gamma)} [\text{meas}(B_r)]^{1/2} < R$. We choose $\delta_1 = \delta_2 = \delta_3 = \delta_s \equiv 0$ and for $i = 1, \ldots, s-1$

$$\delta_{i+3} = \begin{cases} g_i(x^*), & \text{on } B_r, \\ 0, & \text{on } \Gamma \setminus \overline{B_r}. \end{cases}$$

It follows immediately that $\|\delta\|_{Z_0} < R$. For the chosen δ, it is easy to see that x^* is a feasible point for (**LQP**(δ)). Moreover, w^* satisfies the optimality conditions for (5.0.3a)–(5.0.3d). (Note $g_s(x^*) + g_{s,x}(x^*)(x^* - x^*) - \delta_{s+3}$ is zero on B_r and non-positive on $\Gamma \setminus B_r$.) However, we will show that w^* is not only one with respect to the multipliers $(p, \mu_1, \ldots, \mu_s)$, which satisfies (5.0.3a)–(5.0.3d). We choose $\kappa > 0$ and

$$\mu_i = \begin{cases} \kappa, & \text{on } B_r, \\ \mu_i^*, & \text{elsewhere} \end{cases}$$

for all $i = 1, \ldots, s-1$. We set

$$\mu_s = \begin{cases} \mu_s^* + g_{s,u}^{-1}(x^*)(p - p^* - \sum_{i=1}^{s-1} g_{i,u}(x^*)\mu_i), & \text{on } B_r, \\ \mu_s^*, & \text{elsewhere,} \end{cases}$$

where p is the corresponding solution of (5.0.3a). Note that $\mu_i = \mu_i^*$ on $\Gamma \setminus B_r$ for every $i = 1, \ldots, s$, therefore also $p = p^*$ on $\Gamma \setminus B_r$, since w^* is uniquely defined from (5.0.3a)–(5.0.3c) for $\delta \equiv 0$. The tuple $(p, \mu_1, \ldots, \mu_s)$ together

5.1 Stability result for auxiliary problem

with $(y, u) = (y^*, u^*)$ satisfies (5.0.3b)

$$\psi_u(x^*) - p + \sum_{i=1}^{s} g_{i,u}(x^*)\mu_i + \sum_{i=1}^{s} g_{i,uu}(x^*)(u^* - u^*)\mu_i^* + \sum_{i=1}^{s} g_{i,uy}(x^*)(y^* - y^*)\mu_i^*$$

$$= \psi_u(x^*) - p + g_{s,u}(x^*)\mu_s + \sum_{i=1}^{s-1} g_{i,u}(x^*)\mu_i$$

$$= \begin{cases} \psi_u(x^*) - p + g_{s,u}(x^*)\mu_s^* + p - p^* - \sum_{i=1}^{s-1} g_{i,u}(x^*)\mu_i + \sum_{i=1}^{s-1} g_{i,u}(x^*)\kappa, & \text{on } B_r \\ \psi_u(x^*) - p^* + \sum_{i=1}^{s} g_{i,u}(x^*)\mu_i^*, & \text{elsewhere} \end{cases}$$

$$= \psi_u(x^*) - p^* + \sum_{i=1}^{s} g_{i,u}(x^*)\mu_i^* = \delta_2 = 0. \quad \text{(due to the choice of } \delta\text{)}$$

The difference $p - p^*$ satisfies the following adjoint equation

$$a[v, p - p^*] + (d_y(y^*)(p - p^*), v)_\Omega + (b_y(y^*)(p - p^*), v)_\Gamma$$
$$= -\sum_{i=1}^{s}(g_{i,y}(x^*)(\mu_i - \mu_i^*), v)_\Gamma$$

and plugging μ_s yields

$$a[v, p - p^*] + (d_y(y^*)(p - p^*), v)_\Omega + (b_y(y^*)(p - p^*), v)_\Gamma$$
$$= -\left(g_{s,y}(x^*)g_{s,u}^{-1}(x^*)(p - p^* - \sum_{i=1}^{s-1} g_{i,u}(x^*)\mu_i), v\right)_{B_r} - \sum_{i=1}^{s-1}(g_{i,y}(x^*)(\kappa - \mu_i^*), v)_{B_r}$$
$$= -(g_{s,u}^{-1}(x^*)g_{s,y}(x^*)(p - p^*), v)_{B_r} - \sum_{i=1}^{s-1}([g_{i,y}(x^*) - g_{s,u}^{-1}(x^*)g_{s,y}(x^*)g_{i,u}(x^*)]\kappa, v)_{B_r}$$
$$+ \sum_{i=1}^{s-1}(g_{s,u}^{-1}(x^*)g_{s,y}(x^*)\mu_i^*, v)_{B_r}.$$

Note $\mu_s^* \geqslant M > 0$ and $\mu_i = 0$ for all $i = 1, \ldots, s - 1$ on $B_r \subset B \subset A_s$ owing to the separation assumption (**A6**), then the last sum is equal zero on B_r. Since $p = p^*$ on $\Gamma \setminus B_r$ we obtain

$$a[v, p - p^*] + (d_y(y^*)(p - p^*), v)_\Omega + ([b_y(y^*) + g_{s,u}^{-1}(x^*)g_{s,y}(x^*)](p - p^*), v)_\Gamma$$
$$= -\sum_{i=1}^{s-1}([g_{i,y}(x^*) - g_{s,u}^{-1}(x^*)g_{s,y}(x^*)g_{i,u}(x^*)]\kappa, v)_{B_r}.$$

5 Stability of linear-quadratic problems

Moreover, the estimate

$$\|p - p^*\|_{L^\infty(\Gamma)} \leqslant C_{\tau,\infty} \|p - p^*\|_{W^{1,\bar{p}}(\Omega)}$$

$$\leqslant C_{\tau,\infty} C_{p,\Omega} \sum_{i=1}^{s-1} \left\| g_{i,y}(x^*) - g_{s,u}^{-1}(x^*) g_{s,y}(x^*) g_{i,y}(x^*) \right\|_{L^2(\Gamma)} \kappa \, \|\chi_{B_r}\|_{L^2(\Gamma)}$$

$$\leqslant C_{\tau,\infty} C_{p,\Omega} C_2 \sum_{i=1}^{s-1} \left\| g_{i,y}(x^*) - g_{s,u}^{-1}(x^*) g_{s,y}(x^*) g_{i,u}(x^*) \right\|_{L^\infty(\Gamma)} [\operatorname{meas}(\Gamma)]^{1/2}.$$

holds with positive constants C_2 and $C_{\tau,\infty}$ associated to embedding $L^2(\Gamma) \hookrightarrow L^\infty(\Gamma)$ and the trace operator $W^{1,\bar{p}}(\Omega) \to L^\infty(\Gamma)$, respectively, and a positive constant $C_{p,\Omega}$ in the a priori estimate, see Appendix 8.8. By construction $\mu_i \geqslant 0$ for $i = 1, \ldots, s-1$. It remains to show that $\mu_s \geqslant 0$ holds. We find

$$\mu_s = \mu_s^* + g_{s,u}^{-1}(x^*)\bigl(p - p^* - \sum_{i=1}^{s-1} g_{i,u}(x^*)\mu_i\bigr)$$

$$\geqslant M - \left\| g_{s,u}^{-1}(x^*) \right\|_{L^\infty(\Gamma)} \Bigl(\|p - p^*\|_{L^\infty(\Gamma)} + \sum_{i=1}^{s-1} \|g_{i,u}(x^*)\|_{L^\infty(\Gamma)} \kappa \Bigr)$$

$$\geqslant M - \frac{\kappa}{m} \sum_{i=1}^{s-1} l_i \quad \text{on } B_r,$$

where

$$l_i = \|g_{i,u}(x^*)\|_{L^\infty(\Gamma)}$$
$$+ C_{\tau,\infty} C_{p,\Omega} C_2 \left\| g_{i,y}(x^*) - g_{s,u}^{-1}(x^*) g_{s,y}(x^*) g_{i,u}(x^*) \right\|_{L^\infty(\Gamma)} [\operatorname{meas}(\Gamma)]^{1/2}$$

and $\left\| g_{s,u}^{-1}(x^*) \right\|_{L^\infty(\Gamma)} \leqslant 1/m$ thanks to Assumption **(A5)**. Consequently, $\mu_s \geqslant 0$ holds on all of Γ for sufficiently small κ. Therefore, the tuple $(y^*, u^*, p, \mu_1, \ldots, \mu_s)$ satisfies the optimality system (5.0.3a)–(5.0.3d) and is different from w^* in view of $\kappa > 0$. $\square$

In order to show that the solutions of $(\mathbf{LQP^{aux}}(\delta))$ coincide with those of $(\mathbf{LQP}(\delta))$ we need the stability estimates in L^∞. The key is a *projection formula*. [1]

Lemma 5.1.7. *Suppose that Assumptions* **(A1)**-**(A3)**, **(A5)** *and* **(A7)** *hold and that* $\delta \in Z_0$ *and the corresponding* $w_\delta^{aux} \in W_0$ *are given. Then the projec-*

---
[1] This term was firstly used to project the control function u on the interval $[u_a, u_b]$ by studying optimal control-constrained problems ($u_a \leqslant u \leqslant u_b$, see [46]). We apply it to the sum of Lagrange multipliers μ_i.

5.1 Stability result for auxiliary problem

tion formula

$$\sum_{i=1}^{s} g_{i,u}(x^*)\mu_{i,\delta}^{aux} = \max\Big\{0, \delta_2 + p_\delta^{aux} - \psi_u(x^*)$$

$$- \big[\psi_{uy}(x^*) + \sum_{i=1}^{s} g_{i,uy}(x^*)\mu_i^*\big](y_\delta^{aux} - y^*) + \big[\psi_{uu}(x^*) + \sum_{i=1}^{s} g_{i,uu}(x^*)\mu_i^*\big] z\Big\}$$

(5.1.12)

holds a.e. on Γ, where $z = \max_{i=1,\dots,s} g_{i,u}^{-1}(x^*)\big[g_i(x^*) + g_{i,y}(x^*)(y_\delta^{aux} - y^*) - \delta_{i+3}\big]$.

Proof. Owing to **(A5)**, the inequality constraint (5.1.2) yields

$$u_\delta^{aux} - u^* \leqslant -g_{i,u}^{-1}(x^*)\Big(g_i(x^*) + g_{i,y}(x^*)(y_\delta^{aux} - y^*) - \delta_{i+3}\Big) \text{ for all } i=1,\dots,s,$$

$$\Rightarrow u_\delta^{aux} - u^* \leqslant \min_{i=1,\dots,s} -g_{i,u}^{-1}(x^*)\Big(g_i(x^*) + g_{i,y}(x^*)(y_\delta^{aux} - y^*) - \delta_{i+3}\Big)$$

$$= -\max_{i=1,\dots,s} g_{i,u}^{-1}(x^*)\Big(g_i(x^*) + g_{i,y}(x^*)(y_\delta^{aux} - y^*) - \delta_{i+3}\Big).$$

For the gradient equation (5.0.3b), we find

$$\sum_{i=1}^{s} g_{i,u}(x^*)\mu_{i,\delta}^{aux} = \delta_2 + p_\delta^{aux} - \psi_u(x^*) - \psi_{uu}(x^*)(u - u^*) - \psi_{yu}(x^*)(y - y^*)$$

$$- \sum_{i=1}^{s} g_{i,yu}(x^*)(y - y^*)\mu_i^* - \sum_{i=1}^{s} g_{i,uu}(x^*)(u - u^*)\mu_i^*$$

Taking the Legendre-Clebsch condition (Lemma 4.2.2) into account, we obtain that the coefficient before $(u - u^*)$ is negative. Hence we can estimate further

$$\sum_{i=1}^{s} g_{i,u}(x^*)\mu_{i,\delta}^{aux} \geqslant \delta_2 + p_\delta^{aux} - \psi_u(x^*) - \big[\psi_{yu}(x^*) + \sum_{i=1}^{s}\mu_i^* g_{i,yu}(x^*)\big](y_\delta^{aux} - y^*)$$

$$+ \big[\psi_{uu}(x^*) + \sum_{i=1}^{s} g_{i,uu}(x^*)\mu_i^*\big]\max_{i=1,\dots,s} g_{i,u}^{-1}(x^*)\Big(g_i(x^*) + g_{i,y}(x^*)(y_\delta^{aux} - y^*) - \delta_{i+3}\Big).$$

Moreover, by **(A5)** $g_{i,u}(y^*, u^*)$ is positive, and we get

$$\sum_{i=1}^{s} g_{i,u}(x^*)\mu_{i,\delta}^{aux} \geqslant \max\{0, \delta_2 + p_\delta^{aux} - \psi_u(x^*) - \big[\psi_{yu}(x^*) + \sum_{i=1}^{s}\mu_i^* g_{i,yu}(x^*)\big](y_\delta^{aux} - y^*)$$

$$+ \big[\psi_{uu}(x^*) + \sum_{i=1}^{s} g_{i,uu}(x^*)\mu_i^*\big]\max_{i=1,\dots,s} g_{i,u}^{-1}(x^*)\Big(g_i(x^*) + g_{i,y}(x^*)(y_\delta^{aux} - y^*) - \delta_{i+3}\Big)\}.$$

It remains to show that equality holds. On the subset $\Gamma_1 \subset \Gamma$ where the left hand side is strictly positive, we have $\mu_j > 0$ for at least one index j. Therefore,

5 Stability of linear-quadratic problems

the j-th constraint is active, and

$$u_\delta^{aux} - u^* = -\max_{i=1,\ldots,s} g_{i,u}^{-1}(x^*)\Big(g_i(x^*) + g_{i,y}(x^*)(y_\delta^{aux} - y^*) - \delta_{i+3}\Big)$$

holds on Γ_1. On the remainder $\Gamma \setminus \Gamma_1$, the left hand side is zero. Note that $0 \geqslant \max\{0, a\}$ implies that $a \leqslant 0$, and the necessary equality holds. □

The following theorem extends the stability estimate (5.1.9) for auxiliary problem (**LQP^{aux}**(δ)) on W and Z spaces.

Theorem 5.1.8. *Suppose that Assumptions* **(A1)-(A7)** *hold. Then there exists a constant $L^{aux} > 0$ such that*

$$\|w_{\delta'}^{aux} - w_\delta^{aux}\|_W \leqslant L^{aux} \|\delta' - \delta\|_Z \qquad (5.1.13)$$

holds for all $\delta, \delta' \in Z$.

Proof. Let us note that $Z \hookrightarrow Z_0$. We elaborate on the 3D case, i.e., $N = 3$ (the case $N = 2$ is simple). Theorem 5.1.4, the continuity of the trace operator $\tau : H^1(\Omega) \to H^{1/2}(\Gamma)$ and the continuous embedding of $H^{1/2}(\Gamma)$ to $L^4(\Gamma)$ imply that

$$\|\delta y\|_{L^4(\Gamma)} + \|\delta p\|_{L^4(\Gamma)} \leqslant L_4 \|\delta - \delta'\|_Z.$$

The max-operator is Lipschitz continuous from $L^p(\Gamma)$ into itself for $p \in [1, \infty]$. Taking differences of (5.1.12), we obtain

$$\left\|\sum_{i=1}^s g_{i,u}(x^*)\delta\mu_i\right\|_{L^4(\Gamma)} \leqslant L_4' \|\delta - \delta'\|_Z.$$

Thanks to Assumptions **(A5)**–**(A6)** the inequality

$$\|\delta\mu_i\|_{L^4(\Gamma)} \leqslant L_4'' \|\delta - \delta'\|_Z$$

holds for every $i = 1, \ldots, s$. From the gradient equation (5.0.3b) the estimate

$$\|\delta u\|_{L^4(\Gamma)} \leqslant L_4''' \|\delta - \delta'\|_Z$$

follows. Using the same argument as in the proof of Lemma 5.1.2, we can show that the unique solution δy of

$$\mathcal{A}\,\delta y + d_y(y^*)\delta y = r_1 \quad \text{in } \Omega,$$
$$\partial_n \delta y + b_y(y^*)\delta y = r_2 \quad \text{on } \Gamma$$

is an element of $W^{1,\min\{\bar{p},6\}}(\Omega)$ in view of $r_1 = \delta_1 - \delta_1' \in W^{1,\bar{p}'}(\Omega)^*$ and $r_2 = \delta u \in L^4(\Gamma) \hookrightarrow W^{-1/6,6}(\Gamma)$. Similarly, the adjoint equation (5.0.3a) yields the same regularity for δp. To summarize, we get

$$\|\delta y\|_{W^{1,\min\{\bar{p},6\}}(\Omega)} + \|\delta p\|_{W^{1,\min\{\bar{p},6\}}(\Omega)} \leqslant L_4'''' \|\delta - \delta'\|_Z.$$

Since $W^{1,\min\{\bar{p},6\}}(\Omega)$ embeds into $C(\overline{\Omega})$, we obtain the stability for δy and δp also in this space. The projection formula now yields stability for $\delta\mu_i$ in $L^\infty(\Gamma)$, and the gradient equation (5.0.3b) implies the stability of δu in $L^\infty(\Gamma)$. The corresponding improved regularity of r_1 and r_2 now allows to conclude the stability of δy and δp in $W^{1,\bar{p}}(\Omega)$. □

5.2 Relationship between solutions of linear-quadratic and corresponding auxiliary problems

Our goal now is to show that for each δ such that

$$\|\delta\|_Z \leqslant G\,\sigma \tag{5.2.1}$$

with a constant $G > 0$ and $\sigma = \min_i \sigma_i$, the solution $(y_\delta^{aux}, u_\delta^{aux})$ of the auxiliary problem (**LQP**$^{\mathbf{aux}}(\delta)$) coincides with the solution (y_δ, u_δ) of (**LQP**(δ)). Likewise, the Lagrange multipliers and adjoint states of both problems coincide as well. We have seen in Proposition 5.1.6 that the structure of active sets of (**LQP**(δ)) can change dramatically even for arbitrary small perturbations with respect to the Z_0 norms. By contrast, the stability estimate in W with respect to the norm of Z is strong enough in sense that inequality constraints (5.1.2) stay inactive outside of the security sets for small perturbations. This implies that the solutions of (**LQP**$^{\mathbf{aux}}(\delta)$) and (**LQP**(δ)) coincide, see Theorem 5.2.1 below. Let

$$M_\delta := \{x \in X \text{ satisfying (5.0.1) and (5.0.2)}\}$$

be the feasible set for (**LQP**(δ)). Note that $M_\delta \subset M_\delta^{aux}$ for any $\delta \in Z$. The next theorem proves that the set M_δ is non-empty for any such δ, since the solution of (**LQP**$^{\mathbf{aux}}(\delta)$) is also feasible for (**LQP**(δ)). Moreover, following the analogous arguments as in Lemma 5.1.2 there exists a unique solution of (**LQP**(δ)). Therefore, the stability estimate (5.1.13) is valid also for the solution of (**LQP**(δ)).

5 Stability of linear-quadratic problems

Theorem 5.2.1. *There exist $G > 0$ and $L_\delta > 0$ such that $\|\delta\|_Z \leqslant G\sigma$ implies:*

(i) *The solution of* $(\mathbf{LQP^{aux}}(\delta))$ *is feasible for the original problem* $(\mathbf{LQP}(\delta))$.

(ii) *The Lagrange multipliers $(p_\delta, \mu_{1,\delta}, \ldots, \mu_{s,\delta})$ for* $(\mathbf{LQP}(\delta))$ *are unique.*

Moreover, for any δ and δ' satisfying (5.2.1), the corresponding solutions and Lagrange multipliers of $(\mathbf{LQP}(\delta))$ *meet the inequality*

$$\|w_{\delta'} - w_\delta\|_W \leqslant L_\delta \|\delta' - \delta\|_Z. \tag{5.2.2}$$

Proof. Let $x_\delta^{aux} = (y_\delta^{aux}, u_\delta^{aux}) \in M_\delta^{aux}$ be the unique solution of $(\mathbf{LQP^{aux}}(\delta))$. We show that x_δ^{aux} is feasible for $(\mathbf{LQP}(\delta))$, i.e.,

$$g_i(x^*) + g_{i,x}(x^*)(x_\delta^{aux} - x^*) - \delta_{i+3} \leqslant 0 \quad \text{on } \Gamma, \ i = 1, \ldots, s.$$

Since (5.1.2) holds, we have to demonstrate

$$g_i(x^*) + g_{i,x}(x^*)(x_\delta^{aux} - x^*) - \delta_{i+3} \leqslant 0 \quad \text{on } \Gamma \setminus S_i, \ i = 1, \ldots, s.$$

For $\delta = 0$ we have $x_0^{aux} = x^*$. As $g_i(x^*) \leqslant -\sigma_i$ on $\Gamma \setminus S_i$, we estimate

$$g_i(x^*) + g_{i,x}(x^*)(x_\delta^{aux} - x^*) - \delta_{i+3}$$
$$\leqslant -\sigma_i + \|g_{i,x}(x^*)\|_{L^\infty(\Gamma)} \|x^* - x_\delta^{aux}\|_{[L^\infty(\Gamma)]^2} + \|\delta_{i+3}\|_{L^\infty(\Gamma)}.$$

Using (5.1.13) and the trace operator $W^{1,\tilde{p}}(\Omega) \to L^\infty(\Gamma)$ with the constant $C_{\tau,\infty} > 0$ the final estimate

$$g_i(x^*) \leqslant -\sigma_i + (L^{aux} C_{\tau,\infty} \max_{i=1,\ldots,s} \|g_{i,x}(x^*)\|_{L^\infty(\Gamma)} + 1) \|\delta\|_Z \leqslant 0$$

holds a.e. on $\Gamma \setminus S_i^\sigma$ for all $i = 1, \ldots, s$, if

$$\|\delta\|_Z \leqslant \frac{\min_i \sigma_i}{L^{aux} C_{\tau,\infty} \max_i \|g_{i,x}(x^*)\|_{L^\infty(\Gamma)} + 1}.$$

Therewith x_δ^{aux} is feasible for $(\mathbf{LQP}(\delta))$ too and M_δ is nonempty. Moreover, it is closed and convex. We extend the multipliers $(\mu_{1,\delta}^{aux}, \ldots, \mu_{s,\delta}^{aux})$ by zero to all of Γ. Since the objective is a strictly convex function, $(\mathbf{LQP}(\delta))$ possesses a unique solution x_δ and the necessary optimality conditions (5.0.3a)–(5.0.3d) are also sufficient. As w_δ^{aux} satisfies (5.0.3a)–(5.0.3d), then $w_\delta = w_\delta^{aux}$. □

Corollary 5.2.2. *For any $\delta \in Z$ satisfying (5.2.1), we have $\mathcal{A}_i^\delta \subset S_i$ and $\mathcal{A}_j^\delta \subset S_j$, hence $\mathcal{A}_i^\delta \cap \mathcal{A}_j^\delta = \emptyset$ for $i \neq j$.*

Proof. We consider a point $\xi^* \in \mathcal{A}_1^\delta$, i.e.,

$$g_i(x^*(\xi^*)) + g_{i,x}(x^*(\xi^*))(x_\delta(\xi^*) - x^*(\xi^*)) - \delta_{i+3}(\xi^*) = 0$$

holds. We estimate

$$\begin{aligned} g_i(x^*(\xi^*)) &= -g_{i,x}(x^*(\xi^*))(x_\delta(\xi^*) - x^*(\xi^*)) + \delta_{i+3}(\xi^*) \\ &\geq -\max_{i=1,\ldots,s} \|g_{i,x}(x^*)\|_{L^\infty(\Gamma)} \|x^* - x_\delta\|_{[L^\infty(\Gamma)]^2} - \|\delta\|_Z. \end{aligned}$$

Using (5.2.2) and the trace operator $W^{1,\bar{p}}(\Omega) \to L^\infty(\Gamma)$, we find

$$\begin{aligned} g_i(x^*(\xi^*)) &\geq -(L_\delta C_{\tau,\infty} \max_{i=1,\ldots,s} \|g_{i,x}(x^*)\|_{L^\infty(\Gamma)} + 1) \|\delta\|_Z \\ &\geq -\min_{i=1,\ldots,s} \sigma_i \geq -\sigma_i, \ i = 1, \ldots, s. \end{aligned}$$

This shows that $\xi^* \in S_i$ and $\mathcal{A}_i^\delta \subset S_i$ for $i = 1, \ldots, s$. Under Assumption **(A7)** we have $\mathcal{A}_i^\delta \cap \mathcal{A}_j^\delta = \emptyset$, $i \neq j$. $\square$

5.3 Distributed control problem

Within the problem setting (**P'**) the stability result (5.2.2) holds true also for linearized problem (**LQP'**(δ)) with respect to perturbation

$$\delta \in Z' := H_0^1(\Omega) \times L^2(\Omega) \times H_0^1(\Omega) \times [L^2(\Omega)]^s.$$

Let $x^* = (y^*, u^*) \in Y \times L^\infty(\Omega)$ be the optimal solution that together with $(p^*, \mu_1^*, \ldots, \mu_s^*) \in Y \times [L^\infty(\Omega)]^s$ satisfies first and second order optimality conditions. For the further discussion, we abbreviate $w = (y, u, p, \mu_1, \ldots, \mu_s) \in W'$, where

$$W' := Y \times L^\infty(\Omega) \times Y \times [L^\infty(\Omega)]^s.$$

5 Stability of linear-quadratic problems

Analogously to (**LQP**(δ)) and (**LQP**^{aux}(δ)) we define

$$\text{Minimize} \quad J_x(x^*)(x-x^*) + \frac{1}{2}\mathcal{L}_{xx}(w^*)(x-x^*, x-x^*) \qquad (\mathbf{LQP'}(\delta))$$
$$- (\delta_1, y-y^*)_\Omega - (\delta_2, u-u^*)_\Omega$$

subject to $u \in L^\infty(\Omega)$, the linearized state equation

$$\begin{aligned} \mathcal{A}y + d(y^*) + d_y(y^*)(y-y^*) &= u + \delta_3 \quad \text{in } \Omega, \\ y &= 0 \quad \text{on } \Gamma \end{aligned} \qquad (5.3.1)$$

as well as the linearized inequality constraints

$$g_i(x^*) + g_{i,x}(x^*)(x-x^*) \leqslant \delta_{i+3} \qquad \text{a.e. on } \Omega, \qquad (5.3.2)$$
$$i = 1, ..., s.$$

and replacement (5.3.2) by

$$g_i(x^*) + g_{i,x}(x^*)(x-x^*) \leqslant \delta_{i+3} \quad \text{a.e. on } S'_i, \ i = 1, ..., s$$

defines the corresponding auxiliary problem, denoted by (**LQP'** ^{aux}(δ)).

Theorem 5.3.1. *Let Assumptions* (**A1'**)–(**A7'**) *be satisfied. There exist $G' > 0$ and $L'_\delta > 0$ such that*

$$\|\delta\|_{Z'} \leqslant G' \min_i \sigma_i \qquad (5.3.3)$$

implies:

(i) *The solution of* (**LQP'** ^{aux}(δ)) *is feasible for* (**LQP'**(δ)).

(ii) *The Lagrange multipliers* $(p_\delta, \mu_{1,\delta}, \ldots, \mu_{s,\delta})$ *for* (**LQP'**(δ)) *are unique.*

Moreover, for any δ and δ' satisfying (5.3.3), *the corresponding solutions and Lagrange multipliers of* (**LQP'**(δ)) *satisfy the inequality*

$$\|w_{\delta'} - w_\delta\|_{W'} \leqslant L'_\delta \|\delta' - \delta\|_{Z'}.$$

Analogous to Corollary 5.2.2 one shows that the active sets associated with (5.3.2) remain disjoint.

6 Convergence analysis

In this chapter we analyze the local convergence behavior of the SQP method to the optimal solution (y^*, u^*) of (**P**). We define the generalized equation and discuss the relation between SQP and *generalized Newton* (or *Lagrange-Newton*) method. The interplay between these two methods simplifies the convergence analysis. Throughout this chapter, Assumptions (**A1**)–(**A7**) are taken to hold.

6.1 Sequential quadratic programming method

Similarly to the analysis in Section 3.2, the idea of the SQP method also for PDE-constrained optimal control problems is to create the sequence of solutions of quadratic subproblems. More precisely, given an iterate $w^k = (y^k, u^k, p^k, \mu_1^k, \ldots, \mu_s^k)$, one finds the next iterate w^{k+1} as the solution, associated adjoint state and Lagrange multipliers of

$$\text{Minimize} \quad J_x(x^k)(x - x^k) + \frac{1}{2}\mathcal{L}_{xx}(w^k)(x - x^k, x - x^k) \quad (\text{QP}_k)$$

subject to $x \in W^{1,\bar{p}}(\Omega) \times L^\infty(\Gamma)$ and linear state equation

$$\begin{aligned}
\mathcal{A}y + d(y^k) + d_y(y^k)(y - y^k) &= f \quad \text{in } \Omega, \\
\partial_n y + b(y^k) + b_y(y^k)(y - y^k) &= u \quad \text{on } \Gamma
\end{aligned} \quad (6.1.1)$$

and linearized inequality constraints

$$g_i(x^k) + g_{i,x}(x^k)(x - x^k) \leqslant 0 \quad \text{a.e. on } \Gamma, \quad i = 1, \ldots, s. \quad (6.1.2)$$

As we have seen in Section 3.2, the alternative way to compute the same approximate sequence $\{w^k\}$ is to apply Newton's method for the system of optimality posed as an equation or an inclusion [1] in case of the inequality-constrained problem (see (6.2.1) below).

---
[1] later on called *generalized equation*

6.2 Generalized equation

We recall the functional spaces

$$W := W^{1,\bar{p}}(\Omega) \times L^\infty(\Gamma) \times W^{1,\bar{p}}(\Omega) \times [L^\infty(\Gamma)]^s,$$
$$Z := W^{1,\bar{p}'}(\Omega)^* \times L^\infty(\Gamma) \times W^{1,\bar{p}'}(\Omega)^* \times [L^\infty(\Gamma)]^s.$$

and proceed by transforming the optimality conditions (4.1.6a)–(4.1.6d) into a *generalized equation* in the form

$$0 \in F(y, u, p, \mu_1, ..., \mu_s) + N(y, u, p, \mu_1, ..., \mu_s). \tag{6.2.1}$$

For this purpose we define so-called *dual cone* in space $L^\infty(\Gamma)$

$$\mathcal{N}(\mu) := \begin{cases} \{z \in L^\infty(\Gamma) : \langle \nu - \mu, z\rangle \geq 0 \ \forall \nu \in K\}, & \text{if } \mu \in K, \\ \emptyset, & \text{if } \mu \notin K, \end{cases}$$

where $K := \{\nu \in L^\infty(\Gamma) : \nu \geq 0 \text{ a.e. on } \Gamma\}$

and introduce the set-valued mapping $N : W \to 2^Z$ by

$$N(w) := \big(0, 0, 0, \mathcal{N}(\mu_1), ..., \mathcal{N}(\mu_s)\big)^\top,$$

which agrees with the complementarity conditions (4.1.6d). The function $F : W \to Z$ contains the smooth parts and is defined by

$$F(w) = \begin{pmatrix} a[\cdot, p] + (d_y(y)p, \cdot)_\Omega + (\phi_y(y), \cdot)_\Omega \\ + (b_y(y)p, \cdot)_\Gamma + (\psi_y(y, u), \cdot)_\Gamma + \sum_{i=1}^s (g_{i,y}(y, u)\mu_i, \cdot)_\Gamma \\ \psi_u(y, u) - p + \sum_{i=1}^s g_{i,u}(y, u)\mu_i \\ a[y, \cdot] + (d(y), \cdot)_\Omega + (b(y), \cdot)_\Gamma - (f, \cdot)_\Omega - (u, \cdot)_\Gamma \\ g_1(y, u) \\ \cdots \\ g_s(y, u) \end{pmatrix}$$

Lemma 6.2.1. *The first-order necessary conditions (4.1.6a)–(4.1.6d) and the generalized equation (6.2.1) are equivalent.*

Proof. (6.2.1) $\Rightarrow$ (4.1.6a) – (4.1.6d): This is immediate for the first three components. For each component g_i and corresponding Lagrange multiplier

μ_i, $i = 1, ..., s$ we have

$$-g_i(y, u) \in \mathcal{N}_i = \mathcal{N}(\mu_i)$$
$$\Rightarrow \quad \mu_i \in K \quad \text{and} \quad (-g_i(y, u), \nu - \mu_i) \geq 0 \quad \text{for all } \nu \in K$$
$$\Rightarrow \quad \mu_i(\xi) \geq 0 \quad \text{and} \quad -g_i(y(\xi), u(\xi))(\nu - \mu_i(\xi)) \geq 0 \quad \text{for all } \nu \geq 0, \quad \text{a.e. in } \Gamma.$$

This implies
$$\mu_i(\xi) = 0 \quad \Rightarrow \quad g_i(y(\xi), u(\xi)) \leq 0$$
$$\mu_i(\xi) > 0 \quad \Rightarrow \quad g_i(y(\xi), u(\xi)) = 0,$$

which shows the complementarity system in (4.1.6d).

(4.1.6a) – (4.1.6d) $\Rightarrow$ (6.2.1): This is again immediate for the first three components. From the complementarity conditions (4.1.6d) we infer that for every $i = 1, ..., s$

$$g_i(y(\xi), u(\xi)) \nu \leq 0 \quad \text{for all } \nu \geq 0, \quad \text{a.e. in } \Gamma$$
$$\Rightarrow \quad -g_i(y(\xi), u(\xi))(\nu - \mu_i(\xi)) \geq 0 \quad \text{for all } \nu \geq 0, \quad \text{a.e. in } \Gamma$$
$$\Rightarrow \quad -(g_i(y, u), \nu - \mu_i) \geq 0 \quad \text{for all } \nu \in K.$$

In view of $\mu_i \in K$, this implies $-g_i(y, u) \in \mathcal{N}_i = \mathcal{N}(\mu_i)$, $i = 1, ..., s$. $\square$

The Newton method applied to (6.2.1) breaks down the sequence $\{w^k\}$ obtained by solving a linearized generalized equation, i.e., for given w^k the next iterate solves

$$0 \in F(w^k) + F'(w^k)(w - w^k) + N(w). \tag{6.2.2}$$

On the other hand, analogous to Lemma 6.2.1 we can verify that (6.2.2) rep-

6 Convergence analysis

resents the system of optimality for $(\mathbf{QP}_k)$

$$a[v,p] + (d_y(y^k)p, v)_\Omega + (d_{yy}(y^k)(y-y^k)p^k, v)_\Omega + (b_y(y^k)p, v)_\Gamma$$
$$+ (b_{yy}(y^k)(y-y^k)p^k, v)_\Gamma + (\phi_y(y^k), v)_\Omega + (\phi_{yy}(y^k)(y-y^k), v)_\Omega$$
$$+ (\psi_y(y^k, u^k), v)_\Gamma + (\psi_{yy}(y^k, u^k)(y-y^k), v)_\Gamma + (\psi_{yu}(y^k, u^k)(u-u^k), v)_\Gamma$$
$$+ \sum_{i=1}^{s}(g_{i,y}(y^k, u^k)\mu_i, v)_\Gamma + \sum_{i=1}^{s}(g_{i,yy}(y^k, u^k)(y-y^k)\mu_i^k, v)_\Gamma$$
$$+ \sum_{i=1}^{s}(g_{i,yu}(y^k, u^k)(u-u^k)\mu_i^k, v)_\Gamma = 0 \qquad \text{for all } v \in W^{1,\bar{p}'}(\Omega),$$

$$\psi_u(y^k, u^k) + \psi_{uy}(y^k, u^k)(y-y^k) + \psi_{uu}(y^k, u^k)(u-u^k) - p + \sum_{i=1}^{s} g_{i,u}(y^k, u^k)\mu_i$$
$$+ \sum_{i=1}^{s} g_{i,uy}(y^k, u^k)(y-y^k)\mu_i^k + \sum_{i=1}^{s} g_{i,uu}(y^k, u^k)(u-u^k)\mu_i^k = 0, \quad \text{a.e. on } \Gamma,$$

$$a[y,v] + (d(y^k), v)_\Omega + (d_y(y^k)(y-y^k), v)_\Omega + (b(y^k), v)_\Gamma + (b_y(y^k)(y-y^k), v)_\Gamma$$
$$- (f, v)_\Omega - (u, v)_\Gamma = 0 \qquad \text{for all } v \in W^{1,\bar{p}'}(\Omega),$$

$$0 \leqslant \mu_i \perp g_i(x^k) + g_{i,x}(x^k)(x-x^k) \leqslant 0 \qquad \text{a.e. on } \Gamma, \ i=1,\ldots,s.$$

Lemma 6.2.2. *Let x^{k+1} be the solution of $(\mathbf{QP}_k)$ and $(p^{k+1}, \mu_1^{k+1}, \ldots, \mu_s^{k+1})$ be the associated Lagrange multipliers. Then the equation (6.2.2) and the necessary optimality conditions for the minimizer of the quadratic problem $(\mathbf{QP}_k)$ are equivalent.*

Proof. To verify the claim, one follows the proof technique of Lemma 6.2.1. □

6.3 Local convergence result

The main goal of this chapter concerns the local quadratic convergence of the SQP method, described in the previous sections. As was shown in Section 6.2, the SQP method is equivalent to Newton's method (6.2.2), applied to the generalized equation (6.2.1). So it is convenient to carry out the convergence analysis on the level of generalized equations using the implicit function theorem. The local convergence analysis of Newton's method for (6.2.1) is based on a perturbation argument; we refer to [3]. The main ingredients in the proof are the strong regularity of the generalized equation (6.2.1) and local Lipschitz stability of solutions of (6.2.2). We start with the definition of strong regularity.

Let the set $B_r^W(w^*)$ denote an open ball of radius $r > 0$ centered at the point

w^* in the topology of W. With respect to our notions, the strong regularity due to [36] reads as following.

Definition 6.3.1. *The generalized equation (6.2.1) is called <u>strongly regular at point w^*</u> if there exist the radii $r_1 > 0$, $r_2 > 0$ and a positive constant C_L such that for all perturbations $\delta \in B_{r_1}^Z(0)$ the linearized equation*

$$\delta \in F(w^*) + F'(w^*)(w - w^*) + N(w) \qquad (6.3.1)$$

has a unique solution $w_\delta = w(\delta) \in B_{r_2}^W(w^*)$, which satisfies the Lipschitz condition

$$\|w_\delta - w_{\delta'}\|_W \leqslant C_L \|\delta - \delta'\|_Z \qquad \text{for all } \delta, \delta' \in B_{r_1}^Z(0).$$

Thanks to the preparatory results in Chapter 5, this property follows from Theorem 5.2.1.

Lemma 6.3.2. *Under Assumptions (A1)–(A7) the generalized equation (6.2.1) is strong regular at w^*.*

Proof. Since (6.2.1) is equivalent to (4.1.6a)–(4.1.6d) due to Lemma 6.2.1 and Theorem 5.2.1 yields the existence and Lipschitz stability for solution of linear-quadratic subproblem, which lead to existence and stability for solution of the optimality system (4.1.6a)–(4.1.6d), the strong regularity of (6.2.1) follows immediately. □

Let us consider the linearized generalized equation (6.2.2), in which the iterate w^k now plays the role of the parameter η, i.e.,

$$0 \in F(\eta) + F'(\eta)(w - \eta) + N(w) := G(w, \eta) + N(w). \qquad (6.3.2)$$

Since the parameter η presents in (6.3.2) nonlinearly we will apply the implicit function theorem (Theorem 6.3.3 below) to (6.3.2) in order to show the local Lipschitz stability of solutions $w(\eta)$, which is needed in the proof of local quadratic convergence of Newton's step w^k to the optimal solution w^*, see Theorem 6.3.7 below.

First we cite the implicit function theorem from [18, Theorem 2.4] with adapted notation.

Theorem 6.3.3. *Let $w^* \in W$ be a solution to (6.3.2) for $\eta = \eta^*$, let $\mathcal{W}$ be a neighborhood of (w^*, η^*) and $\mathcal{U}$ be a neighborhood of $0 \in Z$. Suppose that*

(i) G is Lipschitz in η, uniformly in w at (w^*, η^*) and $G(w^*, \cdot)$ is Fréchet differentiable at η^* with Fréchet derivative $G_\eta(w^*, \eta^*)\delta\eta$ for all $\delta\eta \in W$;

6 Convergence analysis

(ii) G is partially Fréchet differentiable with respect to w in $\mathcal{W}$ and its partial derivative G_w is continuous in both w and η^* variables at (w^*, η^*);

(iii) (6.3.2) is strongly regular at (w^*, η^*), i.e., there exists a function $\vartheta: \mathcal{U} \to \mathcal{W}$ such that $\vartheta(0) = w^*$, $\delta \in G(w^*, \eta^*) + G_w(w^*, \eta^*)(\vartheta(\delta) - w^*) + N(\vartheta(\delta))$ for all $\delta \in \mathcal{U}$ and Lipschitz continuous ϑ.

Then there exist neighborhoods U of w^* and V of η^* and a function

$$\eta \mapsto w(\eta)$$

from V to U such that $w(\eta)$ is a solution of (6.3.2) for every $\eta \in V$, and $w(\cdot)$ is Lipschitz continuous.

Under Assumption **(A2)**–**(A3)** the condition (ii) of Theorem 6.3.3 holds. Note that (6.3.2) is strongly regular at (w^*, η^*) iff (6.2.1) is strongly regular at w^*, since

$$G(w^*, \eta^*) + G_w(w^*, \eta^*)(\vartheta(\delta) - w^*) + N(\vartheta(\delta))$$
$$= F(\eta^*) + F'(\eta^*)(w^* - \eta^*) + F'(\eta^*)(\vartheta(\delta) - w^*) + N(\vartheta(\delta))$$
$$= F(\eta^*) + F'(\eta^*)(\vartheta(\delta) - \eta^*) + N(\vartheta(\delta)). \quad (w^* = \eta^*)$$

Therefore the assumption (iii) follows from Lemma 6.3.2. It remains to show that G is Lipshitz in η, uniformly in w at (w^*, η^*), i.e., for each (w, η_1) and (w, η_2) in $\mathcal{W}$ there exists a constant $L > 0$ such that

$$\|G(w, \eta_1) - G(w, \eta_2)\|_Z \leqslant L \|\eta_1 - \eta_2\|_W.$$

In other words, to apply Theorem 6.3.3 it remains to verify that the linearized generalized equation (6.3.2) satisfies a Lipschitz condition with respect to η, uniformly in a neighborhood of w^*.

Lemma 6.3.4. *Let Assumptions* **(A1)**–**(A7)** *be valid. There exists the constant* $L > 0$ *and* $r_3 > 0$, $r_4 > 0$ *such that for each* $\eta_1, \eta_2 \in B_{r_3}^W(w^*)$ *and for all* $w \in B_{r_4}^W(w^*)$ *holds*

$$\|F(\eta_1) + F'(\eta_1)(w - \eta_1) - F(\eta_2) - F'(\eta_2)(w - \eta_2)\|_Z \leqslant L \|\eta_1 - \eta_2\|_W.$$

Proof. Let us denote $\eta_i = (y_i, u_i, p_i, \mu_1^i, \ldots, \mu_s^i) \in B_{r_3}^W(w^*)$ and $w = (y, u, p, \mu_1, \ldots, \mu_s) \in B_{r_4}^W(w^*)$, with $r_3, r_4 > 0$ arbitrary. A simple calculation shows

$$F(\eta_1) + F'(\eta_1)(w - \eta_1) - F(\eta_2) - F'(\eta_2)(w - \eta_2)$$

$$= \begin{pmatrix} f_1(\eta_1) - f_1(\eta_2), \\ f_2(\eta_1) - f_2(\eta_2), \\ f_3(y_1) - f_3(y_2), \\ f_4(y_1, u_1) - f_4(y_2, u_2), \\ \ldots \\ f_{s+3}(y_1, u_1) - f_{s+3}(y_2, u_2) \end{pmatrix}$$

where for $i = \{1, 2\}$

$$f_1(\eta_i) = (d_y(y_i) \, p + \phi_y(y_i) + [\phi_{yy}(y_i) + d_{yy}(y_i) \, p_i](y - y_i), \cdot)_\Omega$$
$$+ (b_y(y_i) \, p + \psi_y(y_i, u_i) + \sum_{j=1}^{s} g_{j,y}(y_i, u_i) \, \mu_j, \cdot)_\Gamma$$
$$+ (\psi_{yy}(y_i, u_i) + b_{yy}(y_i) \, p_i + \sum_{j=1}^{s} g_{j,yy}(y_i, u_i) \, \mu_j^i, y - y_i)_\Gamma$$
$$+ (\psi_{yu}(y_i, u_i) + \sum_{j=1}^{s} g_{j,yu}(y_i, u_i) \, \mu_j^i, u - u_i)_\Gamma,$$

$$f_2(\eta_i) = \psi_u(y_i, u_i) + \psi_{uy}(y_i, u_i)(y - y_i) + \psi_{uu}(y_i, u_i)(u - u_i)$$
$$+ \sum_{j=1}^{s} g_{j,u}(y_i, u_i) \, \mu_j + \sum_{j=1}^{s} g_{j,yu}(y_i, u_i) \, \mu_j^i \, (y - y_i)$$
$$+ \sum_{j=1}^{s} g_{j,uu}(y_i, u_i) \, \mu_j^i (u - u_i),$$

$$f_3(y_i) = (d(y_i) + d_y(y_i)(y - y_i), \cdot)_\Omega + (b(y_i) + b_y(y_i)(y - y_i), \cdot)_\Gamma,$$
$$f_{3+j}(y_i, u_i) = g_j(y_i, u_i) + g_{j,y}(y_i, u_i)(y - y_i) + g_{j,u}(y_i, u_i), \qquad j = 1, \ldots, s.$$

We examine only the Lipschitz condition for f_3; the rest follows analogously. That means we have to estimate

$$\|f_3(y_1) - f_3(y_2)\|_{W^{1,\bar{p}'}(\Omega)^*} = \sup_{\|v\|_{W^{1,\bar{p}'}(\Omega)}=1} |\langle f_3(y_1) - f_3(y_2), v \rangle_{W^{1,\bar{p}'}(\Omega)^*, W^{1,\bar{p}'}(\Omega)}|.$$

6 Convergence analysis

We consider

$$|\langle f_3(y_1) - f_3(y_2), v\rangle_{W^{1,\bar{p}'}(\Omega)^*, W^{1,\bar{p}'}(\Omega)}|$$
$$\leqslant |(d(y_1) - d(y_2) + d_y(y_1)(y - y_1) - d_y(y_2)(y - y_2), v)_\Omega|$$
$$+ |(b(y_1) - b(y_2) + b_y(y_1)(y - y_1) - b_y(y_2)(y - y_2), v)_\Gamma|$$
$$\leqslant \|d(y_1) - d(y_2) + d_y(y_1)(y - y_1) - d_y(y_2)(y - y_2)\|_{L^2(\Omega)} \|v\|_{L^2(\Omega)}$$
$$+ \|b(y_1) - b(y_2) + b_y(y_1)(y - y_1) - b_y(y_2)(y - y_2)\|_{L^2(\Gamma)} \|v\|_{L^2(\Gamma)}$$

Using the triangle inequality and embeddings $W^{1,\bar{p}'}(\Omega) \hookrightarrow L^2(\Omega)$ and $W^{1,\bar{p}'}(\Omega) \hookrightarrow L^2(\Gamma)$, we estimate first summand by

$$C_2 \big[\|d(y_1) - d(y_2)\|_{L^2(\Omega)} + \|d_y(y_1)(y_2 - y_1)\|_{L^2(\Omega)}$$
$$+ \|(d_y(y_1) - d_y(y_2))(y - y_2)\|_{L^2(\Omega)} \big] \|v\|_{W^{1,\bar{p}'}(\Omega)}$$

and the second summand by

$$C_2' \big[\|b(y_1) - b(y_2)\|_{L^2(\Gamma)} + \|b_y(y_1)(y_2 - y_1)\|_{L^2(\Gamma)}$$
$$+ \|(b_y(y_1) - b_y(y_2))(y - y_2)\|_{L^2(\Gamma)} \big] \|v\|_{W^{1,\bar{p}'}(\Omega)},$$

where $C_2 > 0$ and $C_2' > 0$ are corresponding embedding constants. Note

$$\|d_y(y_1)(y_2 - y_1)\|_{L^2(\Omega)} \leqslant \|d_y(y_1)\|_{L^\infty(\Omega)} \|y_2 - y_1\|_{L^2(\Omega)},$$
$$\|(d_y(y_1) - d_y(y_2))(y - y_2)\|_{L^2(\Omega)} \leqslant \|d_y(y_1) - d_y(y_2)\|_{L^\infty(\Omega)} \|y - y_2\|_{L^2(\Omega)}.$$

The properties of d, see Lemma 8.6.4, imply that $\|d_y(y_1)\|_{L^\infty(\Omega)}$ is uniformly bounded for all $y_1 \in B_{r_3}^{L^\infty(\Omega)}(y^*)$. Moreover,

$$\|y - y_2\|_{L^2(\Omega)} \leqslant \|y - y^*\|_{L^2(\Omega)} + \|y^* - y_2\|_{L^2(\Omega)} \leqslant C_2 (r_3 + r_4).$$

Together with the Lipschitz properties of d and d_y, see again Lemma 8.6.4 we obtain

$$\|d(y_1) - d(y_2)\|_{L^2(\Omega)} + \|d_y(y_1)(y_2 - y_1)\|_{L^2(\Omega)} + \|(d_y(y_1) - d_y(y_2))(y - y_2)\|_{L^2(\Omega)}$$
$$\leqslant L_1 \|y_2 - y_1\|_{L^2(\Omega)} \leqslant L_1 C_2 \|y_2 - y_1\|_{W^{1,\bar{p}}(\Omega)}$$

for some constant $L_1 > 0$. The analogous arguments yield the estimate

$$\|b(y_1) - b(y_2)\|_{L^2(\Gamma)} + \|b_y(y_1)(y_2 - y_1)\|_{L^2(\Gamma)} + \|(b_y(y_1) - b_y(y_2))(y - y_2)\|_{L^2(\Gamma)}$$
$$\leqslant L_2 \|y_2 - y_1\|_{W^{1,\bar{p}}(\Omega)}.$$

with a positive constant L_2. Thus

$$|\langle f_3(y_1) - f_3(y_2), v\rangle_{W^{1,\bar{p}'}(\Omega)^*, W^{1,\bar{p}'}(\Omega)}| \leqslant L \, \|y_1 - y_2\|_{W^{1,\bar{p}}(\Omega)} \, \|v\|_{W^{1,\bar{p}'}(\Omega)}$$

holds for some constant $L > 0$. The last inequality implies

$$\|f_3(y_1) - f_3(y_2)\|_{W^{1,\bar{p}'}(\Omega)^*} \leqslant L \, \|y_1 - y_2\|_{W^{1,\bar{p}}(\Omega)}.$$

$\square$

Now the existence and Lipschitz continuity for the solution $w = w(\eta)$ of (6.3.2) with respect to η follows from Theorem 6.3.3 directly.

Theorem 6.3.5. *Suppose that Assumptions* **(A1)**-**(A7)** *hold. There exist radii* r_5, $r_6 > 0$ *such that for each* $\eta_1, \eta_2 \in B_{r_5}^W(w^*)$ *the corresponding solutions* $w(\eta_i) \in B_{r_6}^W(w^*)$, $i = 1, 2$ *and the inequality*

$$\|w(\eta_1) - w(\eta_2)\|_W \leqslant L_\eta \|\eta_1 - \eta_2\|_W$$

holds with some constant $L_\eta > 0$.

Moreover, we can prove that for each $w^k \in B_R^{W^\infty}(w^*)$ there exists a unique global solution of $(\mathbf{QP}_k)$. $(W^\infty = L^\infty(\Omega) \times L^\infty(\Gamma) \times L^\infty(\Omega) \times [L^\infty(\Gamma)]^s)$

Lemma 6.3.6. *Suppose Assumptions* **(A1)**–**(A7)** *are satisfied. Let* $w^k \in W$ *be fixed. There exists* $R > 0$ *such that* $(\mathbf{QP}_k)$ *has a unique global solution* $x = (y, u) \in X$, *provided that* $w^k \in B_R^{W^\infty}(w^*)$.

Proof. We define the set of admissible points for $(\mathbf{QP}_k)$ by

$$M^k := \{x \in X \text{ satisfying (6.1.1) and (6.1.2)}\}.$$

The set M^k depends on the parameter x^k but for each fixed parameter is

- convex, since the equality and inequality constraints in $(\mathbf{QP}_k)$ are linear with respect to the variable $x = (y, u) \in X$,

- closed, and

- nonempty.
  Due to Theorem 6.3.3 and Lemma 6.3.2 for any $x^k \in X$ there exists a locally unique solution of the linearized generalized equation (6.2.2), which is equivalent to the system of optimality for the solution of $(\mathbf{QP}_k)$. Thus, (6.1.1) and (6.1.2) are satisfied and M^k admits a feasible pair (y, u).

6 Convergence analysis

The objective of $(\mathbf{QP}_k)$ can be decomposed into quadratic and affine parts in x, analogous to how it was done in Lemma 5.1.2. Thus it satisfies the quadratic growth condition, since for each fixed pair (y^k, u^k)

$$\exists\, \alpha'' > 0: \quad \mathcal{L}''_{xx}(x^k, \lambda^k)(x - x^k, x - x^k) \geqslant \alpha'' \left\| x - x^k \right\|^2_{[L^2(\Omega)]^2}$$

holds for all $x = (y, u) \in X$ satisfying the linearized PDE

$$\mathcal{A}y + d(y^k) + d_y(y^k)(y - y^k) = 0 \quad \text{in } \Omega$$
$$\partial_n y + b(y^k) + b_y(y^k)(y - y^k) = u - u^k \quad \text{on } \Gamma$$

and provided that $\left\| w^k - w^* \right\|_{W^\infty} < R$ due to Lemma 4.2.4. Thanks to Theorem 8.1.5 the problem $(\mathbf{QP}_k)$ has a unique solution (y, u). $\square$

Theorem 6.3.7 below summarizes the convergence properties of the Lagrange-Newton method as well as of the SQP method, due to their equivalence.

Theorem 6.3.7. *There exists a radius $r > 0$ and a constant $C_{SQP} > 0$ such that for each starting point $w^0 \in B_r^W(w^*)$, the sequence of iterates w^k generated by (6.2.2) is well-defined in $B_r^W(w^*)$ and satisfy*

$$\left\| w^{k+1} - w^* \right\|_W \leqslant C_{SQP} \left\| w^k - w^* \right\|^2_W. \tag{6.3.3}$$

Proof. Suppose that the iterate $w^k \in B_r^W(w^*)$ is given. The radius r satisfying $r_5 \geqslant r > 0$ will be specified below. From Theorem 6.3.5, we infer the existence of a solution w^{k+1} of (6.2.2) which is unique in $B_{r_6}^W(w^*)$. That is, we have

$$0 \in F(w^*) + F'(w^*)(w^* - w^*) + N(w^*), \tag{6.3.4a}$$
$$0 \in F(w^k) + F'(w^k)(w^{k+1} - w^k) + N(w^{k+1}). \tag{6.3.4b}$$

Adding and subtracting the terms $F'(w^*)(w^{k+1} - w^*)$ and $F(w^*)$ to (6.3.4b), we obtain

$$\delta^{k+1} \in F(w^*) + F'(w^*)(w^{k+1} - w^*) + N(w^{k+1}) \tag{6.3.5}$$

where

$$\delta^{k+1} := F(w^*) - F(w^k) + F'(w^*)(w^{k+1} - w^*) - F'(w^k)(w^{k+1} - w^k).$$

From Lemma 6.3.4 with $\eta_1 := w^*$, $\eta_2 := w^k$, $w := w^{k+1}$, and $r_3 := r_5$, $r_4 := r_6$, we get

$$\left\| \delta^{k+1} \right\|_Z \leqslant L \left\| w^k - w^* \right\|_W < L\, r, \tag{6.3.6}$$

where L depends only on the radii. That is, $\|\delta^{k+1}\|_Z \leq G\sigma$ holds whenever

$$r \leq \frac{G\sigma}{L},$$

which we impose on r.

Analogous to Lemma 6.2.2 one can show that (6.3.4a) and (6.3.5) are equivalent to the system of optimality corresponding to problem $(\mathbf{LQP}(\delta))$ for $\delta = 0$ and $\delta = \delta^{k+1}$, respectively. From Theorem 5.2.1, we thus obtain

$$\|w^{k+1} - w^*\|_W \leq L_\delta \|\delta^{k+1} - 0\|_Z. \tag{6.3.7}$$

It remains to verify that $\|\delta^{k+1}\|_Z$ is quadratic in $\|w^k - w^*\|_W$. We estimate

$$\|\delta^{k+1}\|_Z \leq \|F(w^*) - F(w^k) - F'(w^k)(w^* - w^k)\|_Z$$
$$+ \|(F'(w^*) - F'(w^k))(w^{k+1} - w^*)\|_Z.$$

As in the proof of Theorem 4.2.5, the first term is bounded by a constant times $\|w^k - w^*\|^2_{L^\infty(\Omega)}$. Moreover, the Lipschitz properties of the terms in F' imply that the second term is bounded by a constant times $\|w^k - w^*\|_{L^\infty(\Omega)} \|w^{k+1} - w^*\|_{L^2(\Omega)}$. We thus conclude

$$\|\delta^{k+1}\|_Z \leq c_1 \|w^k - w^*\|^2_W + c_2 \|w^k - w^*\|_W \|w^{k+1} - w^*\|_W, \tag{6.3.8}$$

where the constants depend only on the radius r_5. We finally choose r as

$$r = \min\left\{r_5, \frac{G\sigma}{L}, \frac{1}{L_\delta \max\{2 c_2, c_1 + c_2 L_\delta L\}}\right\}.$$

Then (6.3.6)–(6.3.8) imply $w^{k+1} \in B_r^W(w^*)$ since

$$\|w^{k+1} - w^*\|_W < L_\delta [c_1 r + c_2 \|w^{k+1} - w^*\|_W] r$$
$$\leq L_\delta [c_1 + c_2 L_\delta L] r^2 \leq r.$$

Moreover, (6.3.7)–(6.3.8) yield

$$\|w^{k+1} - w^*\|_W \leq L_\delta c_1 \|w^k - w^*\|^2_W + c_2 L_\delta r \|w^{k+1} - w^*\|_W$$

and thus

$$\|w^{k+1} - w^*\|_W \leq C_{SQP} \|w^k - w^*\|^2_W$$

holds with $C_{SQP} = \frac{L_\delta c_1}{1 - c_2 L_\delta r}$. □

6 Convergence analysis

Clearly, Theorem 6.3.7 proves the local quadratic convergence of the SQP method. Recall that the iterates w^k are defined by means of Lemma 6.3.6, as the *local unique* solutions, Lagrange multipliers and adjoint states of $(\mathbf{QP}_k)$. Next we show that by the SQP method under Assumption **(A6)** all iterates stay in the neighborhood of w^*. Indeed, we define a radius r' such that for all $w^k \in B_{r'}^W(w^*)$

- x^{k+1} is *globally unique* solution of $(\mathbf{QP}_k)$,
- the active sets $\mathcal{A}_k$ stay disjoint,
- $(p^{k+1}, \mu_1^{k+1}, \ldots, \mu_s^{k+1})$ are unique Lagrange multipliers associated to x^{k+1}.

Corollary 6.3.8. *There exists a radius $r' > 0$ such that $w^k \in B_{r'}^W(w^*)$ implies that $(\mathbf{QP}_k)$ has a unique global solution x^{k+1}. The associated Lagrange multipliers and adjoint state $\lambda^{k+1} = (p^{k+1}, \mu_1^{k+1}, \ldots, \mu_s^{k+1})$ are also unique. The iterate w^{k+1} lies again in $B_{r'}^W(w^*)$.*

Proof. We first observe that Theorem 6.3.7 remains valid (with the same constant C_{SQP}) if r is taken to be smaller than chosen in the proof. Here, we set

$$r' = \min\left\{\sigma, \frac{\sigma}{3\,C_{bound}}, R, r\right\},$$

where R and r are the radii from Lemma 6.3.6 and Theorem 6.3.7, respectively, and $C_{bound} = \max\{L_g, \max_i \|g_{i,x}(x)\|_{L^\infty(\Gamma)}\}$ and L_g is a maximal Lipschitz constant for $g_i(y, u)$.

Suppose that $w^k \in B_{r'}^W(w^*)$ holds. Then Lemma 6.3.6 implies that $(\mathbf{QP}_k)$ possesses a globally unique solution $x^{k+1} \in X$. The corresponding active sets are defined by

$$\mathcal{A}_i^{k+1} := \{\xi \in \Gamma : g_i(x^k) + g_{i,x}(x^k)(x^{k+1} - x^k) = 0\}, \ i = 1, \ldots, s.$$

We show that $\mathcal{A}_i^{k+1} \subset S_i$. For almost every $\xi \in \mathcal{A}_i^{k+1}$, we have

$$\begin{aligned}
g_i(\xi, x^*(\xi)) &= g_i(\xi, x^*(\xi)) - g_i(\xi, x^k(\xi)) + g_{i,x}(\xi, x^k(\xi))(x^k(\xi) - x^{k+1}(\xi)) \\
&\geq -\left(L_g + \|g_{i,x}(x^k)\|_{L^\infty(\Gamma)}\right)\|x^* - x^k\|_{L^\infty(\Omega) \times L^\infty(\Gamma)} \\
&\quad - \|g_{i,x}(x^k)\|_{L^\infty(\Gamma)} \|x^* - x^{k+1}\|_{L^\infty(\Omega) \times L^\infty(\Gamma)} \\
&\geq -\left(L_g + \max_i \|g_{i,x}(x^k)\|_{L^\infty(\Gamma)}\right)\|x^* - x^k\|_{L^\infty(\Omega) \times L^\infty(\Gamma)} \\
&\quad - \max_i \|g_{i,x}(x^k)\|_{L^\infty(\Gamma)} \|x^* - x^{k+1}\|_{L^\infty(\Omega) \times L^\infty(\Gamma)} \\
&\geq -\left(L_g + 2\max_i \|g_{i,x}(x^k)\|_{L^\infty(\Gamma)}\right) r' \\
&\geq -3\,C_{bound} \cdot r' \geq -\min_i \sigma_i \geq -\sigma_i, \quad i = 1, \ldots, s,
\end{aligned}$$

since Theorem 6.3.7 implies that $w^{k+1} \in B_{r'}^W(w^*)$ and thus in particular $x^{k+1} \in B_{r'}^X(x^*)$. Owing to Assumption **(A6)**, the active sets $\mathcal{A}_i^{k+1}$, $i = 1, ..., s$ are disjoint. Therefore analogous to Theorem 4.1.6 the Lagrange multipliers μ_i^{k+1}, $i = 1, ..., s$ and adjoint state p^{k+1} can be uniquely defined. □

Thus the SQP method which is equivalent to Newton's approach yields a sequence $\{w^k\}$, which converges quadratically according to (6.3.3), if the initial point is chosen close enough to w^*.

6.4 Distributed control problem

In a similar fashion, we prove the local quadratic convergence of the SQP method applied to the distributed optimal control problem **(P')**. One only needs to make the necessary adaptation with respect to the problem setting. Since for **(P')** $u \in L^\infty(\Omega)$ and the inequality constraints (4.3.2) are given in Ω, the generalized equation reads as

$$0 \in F(w) + N(w) \quad (6.4.1)$$

where $F : W' \to Z'$ and $N : W' \to 2^{Z'}$ is a set-valued function, which contains the dual cones in $L^\infty(\Omega)$.

Definition 6.4.1. The <u>dual cone</u> $\mathcal{N}'(\mu)$ in space $L^\infty(\Omega)$ is defined by

$$\mathcal{N}'(\mu) := \begin{cases} \{z \in L^\infty(\Omega) : \langle \mu - \nu, z \rangle \leqslant 0 \ \forall \nu \in K'\}, & \mu \in K', \\ \emptyset, & \mu \notin K', \end{cases}$$

where $K' := \{\mu \in L^\infty(\Omega) : \mu \geqslant 0 \text{ a.e. in } \Omega\}$.

Using the result of Theorem 5.3.1 we can show that the generalized equation is strong regular at $w^* \in W'$. Similarly, the implicit function theorem provides the existence and Lipschitz continuity for the solution $w = w(w^k)$ of the linearized generalized equation

$$0 \in F(w^k) + F'(w^k)(w - w^k) + N(w) \quad (6.4.2)$$

with respect to w^k considered as a parameter. Therefore, the Newton's method applied to (6.4.1) (or the SQP method applied to **(P')**) yields the local quadratic convergence of iterate $\{w^k\}$ to the optimal solution w^*, see Theorems 6.4.2–6.4.3 below.

6 Convergence analysis

Theorem 6.4.2. *Let Assumptions* (**A1'**)–(**A7'**) *be satisfied. Then the generalized equation* (6.2.1) *is strong regular at the point* $w^* \in W'$.

Theorem 6.4.3. *Under Assumptions* (**A1'**)–(**A7'**), *there exist a radius* $r' > 0$ *and a constant* $C'_{SQP} > 0$ *such that for each starting point* $w^0 \in B_{r'}^{W'}(w^*)$, *the sequence of iterates* w^k *generated by* (6.4.2) *is well-defined in* $B_{r'}^{W'}(w^*)$ *and satisfy*

$$\left\|w^{k+1} - w^*\right\|_{W'} \leqslant C'_{SQP} \left\|w^k - w^*\right\|_{W'}^2. \tag{6.4.3}$$

7 Numerical realization

This chapter deals with the numerical realization of the SQP method discussed in Chapter 6. Because of the presence both an elliptic PDE and inequality constraints, the practical implementation acts with discretization technique and primal-dual active set strategy. Therefore to be able to realize numerical computations we adopt the SQP algorithm with respect to both discretization and PDAS methods. Numerical tests are illustrated in Section 7.3.

7.1 Primal-Dual Active Set Strategy

The primal-dual active set (PDAS) method has proved to be an efficient numerical tool in the context of optimization problems including inequality constraints, e.g. [9]. PDAS strategies are composed in two major components: determination of the active sets, and computation of the solutions of optimal control subproblems: constrained on these active sets and unconstrained outside. More precisely, in each step of the iteration method, a set of active constraints is defined. On active sets, the inequality constraints perform equalities. Therefore, the associated parts of the control and state can be defined from the optimization problem, in which inequality constraints are replaced by corresponding equalities. Its remaining parts are obtained from the optimality system for the associated unconstrained problem.

In the SQP method we shall apply the PDAS procedure to the linear-quadratic subproblem $(\mathbf{QP}_k)$, which we recall:

$$\text{Minimize} \quad J_x(x^k)(x - x^k) + \frac{1}{2}\mathcal{L}_{xx}(w^k)(x - x^k, x - x^k)$$

subject to $x \in W^{1,\bar{p}}(\Omega) \times L^\infty(\Gamma)$ and linear state equation

$$\mathcal{A}y + d(y^k) + d_y(y^k)(y - y^k) = f \quad \text{in } \Omega,$$
$$\partial_n y + b(y^k) + b_y(y^k)(y - y^k) = u \quad \text{on } \Gamma$$

and linearized inequality constraints

$$g_i(x^k) + g_{i,x}(x^k)(x - x^k) \leqslant 0 \quad \text{a.e. on } \Gamma, \quad i = 1, ..., s.$$

7 Numerical realization

Complementarity conditions for $(\mathbf{QP}_k)$

$$0 \leqslant \mu_i \perp g_i(x^k) + g_{i,x}(x^k)(x - x^k) \leqslant 0 \quad \text{a.e. on } \Gamma, \ i = 1, \ldots, s \qquad (7.1.1)$$

play a fundamental role in PDAS strategy, since they characterize the active set $\mathcal{A}_i^k$ as well as an associated multiplier μ_i^k. The next lemma displays various equivalent representations of (7.1.1).

Lemma 7.1.1. *Let $x = (y, u) \in X$ be the solution of $(\mathbf{QP}_k)$ and $\mu_i \in L^\infty(\Gamma)$ be Lagrange multipliers. Then the following statements are equivalent:*

(i) *the complementarity conditions (7.1.1) are valid;*

(ii) *for $i = 1, \ldots, s$*

$$\begin{cases} g_i(x^k) + g_{i,x}(x^k)(x - x^k) = 0 & \text{if } \mu_i > 0, \\ g_i(x^k) + g_{i,x}(x^k)(x - x^k) \leqslant 0 & \text{if } \mu_i = 0; \end{cases}$$

(iii) *for $i = 1, \ldots, s$ and any $c_i > 0$*

$$\mu_i = \max\left\{0, \mu_i + c_i[g_i(x^k) + g_{i,x}(x^k)(x - x^k)]\right\}. \qquad (7.1.2)$$

Proof. The proof is realized by three steps.

(i) $\Rightarrow$ (ii) We rewrite (7.1.1) explicitly: $\mu_i \geqslant 0$, $g_i(x^k) + g_{i,x}(x^k)(x - x^k) \leqslant 0$ and $\mu_i[g_i(x^k) + g_{i,x}(x^k)(x - x^k)] = 0$. The claim is clear now.

(ii) $\Rightarrow$ (iii) If $g_i(x^k) + g_{i,x}(x^k)(x - x^k) \leqslant 0$ and $\mu_i = 0$, then

$$\mu_i = \max\{0, \mu_i + c_i[g_i(x^k) + g_{i,x}(x^k)(x - x^k)]\}.$$

If $g_i(x^k) + g_{i,x}(x^k)(x - x^k) = 0$ and $\mu_i > 0$, then

$$\mu_i = \max\{0, \mu_i\} = \max\{0, \mu_i + c_i[g_i(x^k) + g_{i,x}(x^k)(x - x^k)]\}.$$

(iii) $\Rightarrow$ (i) Let $\mu_i = \max\{0, \mu_i + c_i[g_i(x^k) + g_{i,x}(x^k)(x - x^k)]\}$. Then

a) $\mu_i = 0 \Rightarrow \mu_i + c_i[g_i(x^k) + g_{i,x}(x^k)(x - x^k)] \leqslant 0$
$\Rightarrow g_i(x^k) + g_{i,x}(x^k)(x - x^k) \leqslant 0$, since $c_i > 0$.

b) $\mu_i > 0 \Rightarrow \mu_i = \mu_i + c_i[g_i(x^k) + g_{i,x}(x^k)(x - x^k)]$
$\Rightarrow g_i(x^k) + g_{i,x}(x^k)(x - x^k) = 0$, since $c_i > 0$.

In both cases, $\mu_i[g_i(x^k) + g_{i,x}(x^k)(x - x^k)] = 0$ is satisfied. Thus (7.1.1) holds.

7.1 Primal-Dual Active Set Strategy

□

Analogously, the non-smooth representations of the complementary conditions (4.1.6d) are

$$\mu_i = \max\{0, \mu_i + c_i g_i(x)\}, \quad i = 1, \ldots, s. \tag{7.1.3}$$

Using (7.1.2) the PDAS algorithm for $(\mathbf{QP}_k)$ reads as follows.

Algorithm 7.1.2.

1. **set** $n = 0$ *and* **choose** *the initial value* w_0

2. **define** *the active sets for fixed* x^k

$$\mathcal{A}_{i,n}^k := \{\xi \in \Gamma : \ \mu_{i,n} + c_i[g_i(\xi, x^k(\xi)) + g_{i,x}(\xi, x^k(\xi))(x_n(\xi) - x^k(\xi))] > 0\}$$
$$\mathcal{I}_{i,n}^k := \Gamma \setminus \mathcal{A}_{i,n}^k$$

3. **substitute** $(\mathbf{QP}_k)$ *by*

$$\text{Minimize} \quad J_x(x^k)(x - x^k) + \frac{1}{2}\mathcal{L}_{xx}(w^k)(x - x^k, x - x^k) \quad (\mathbf{QP}_{\mathcal{A}_{i,n}^k})$$

subject to $u \in L^\infty(\Gamma)$, *the linearized elliptic equation*

$$\mathcal{A}y + d(y^k) + d_y(y^k)(y - y^k) = f \quad \text{in } \Omega,$$
$$\partial_n y + b(y^k) + b_y(y^k)(y - y^k) = u \quad \text{on } \Gamma$$

and linearized equality constraints

$$g_i(x^k) + g_{i,x}(x^k)(x - x^k) = 0 \text{ a.e. on } \mathcal{A}_{i,n}^k.$$

4. **compute** $w_{n+1} = (y_{n+1}, u_{n+1}, p_{n+1}, \mu_{1,n+1}, \ldots, \mu_{s,n+1})$

5. **examine** *the stopping criterion*
  (STOP or GOTO step 2. with $n := n + 1$*)*

Note the following cases may occur:

- the active sets $\mathcal{A}_{i,n}^k$ intersect and multipliers $(p_{n+1}, \mu_{1,n+1}, \ldots, \mu_{s,n+1})$ may be non-unique (see Example 4.1.4);

- there exists no feasible point for $(\mathbf{QP}_{\mathcal{A}_{i,n}^k})$;

- $(\mathbf{QP}_{\mathcal{A}_{i,n}^k})$ is indefinite.

7 Numerical realization

Due to Corollary 6.3.8 if w^k is located in the vicinity of w^*, then the active sets $\mathcal{A}_i^k$ stay disjoint under Assumption **(A7)**. Since the active sets $\mathcal{A}_i^k$ do not intersect, we can restrict ourself to the case where $\mathcal{A}_{i,n}^k$ are disjoint for every n. Therefore Algorithm 7.1.2 provides the unique solution, if w^k is sufficiently close to the solution w^*. Thus, one needs to specify the security sets S_i according to Definition 4.1.5, ensure that w^k be close to w^* and use that $\mathcal{A}_{i,n}^k$ are disjoint for each n. The verification of these issues supposes w^* is already known, which is usually not available for real-life problems. A remedy is the penalization strategy applied to the inequality constraints, i.e., one solves $(\mathbf{QP}_k)$ by treating the i-th inequality constraint on the corresponding active set $\mathcal{A}_i^k$ by PDAS and penalizing the other constraints by adding the terms $\gamma_j \left\| [g_j(x^k) + g_{j,x}(x^k)(x - x^k)]^+ \right\|_{L^2(\Gamma)}^2$, $j \neq i$ to the objective function of $(\mathbf{QP}_k)$.

In the following, we suppose Assumptions **(A1)–(A7)** are satisfied. Thus $(\mathbf{QP}_k)$ is well-defined and $\mathcal{A}_i^k$ stay disjoint for each iteration k. Therefore, $\mathcal{A}_{i,n}^k$ can be defined using following nested procedure:

- we define the active $\mathcal{A}_{1,n}^k$ and inactive $\mathcal{I}_{1,n}^k$ sets according to the first inequality

$$\mathcal{A}_{1,n}^k := \{\xi \in \Gamma : \mu_1 + c_1[g_1(\xi, x^k(\xi)) + g_{1,x}(\xi, x^k(\xi))(x_n(\xi) - x^k(\xi))] > 0\},$$
$$\mathcal{I}_{1,n}^k := \Gamma \setminus \mathcal{A}_{1,n}^k;$$

- with fixed $\mathcal{A}_{1,n}^k$ and $\mathcal{I}_{1,n}^k$, we determine the active $\mathcal{A}_{2,n}^k$ and inactive $\mathcal{I}_{2,n}^k$ sets associated with the second inequality

$$\mathcal{A}_{2,n}^k := \{\xi \in \mathcal{I}_1^k : \mu_2 + c_2[g_2(\xi, x^k(\xi)) + g_{2,x}(\xi, x^k(\xi))(x_n(\xi) - x^k(\xi))] > 0\},$$
$$\mathcal{I}_{2,n}^k := \Gamma \setminus (\mathcal{A}_{1,n}^k \cup \mathcal{A}_{2,n}^k);$$

- the sets $\mathcal{A}_{i,n}^k$ and $\mathcal{I}_{i,n}^k$ one defines with fixed $\mathcal{A}_{1,n}^k, \ldots, \mathcal{A}_{i-1,n}^k$ and $\mathcal{I}_{1,n}^k, \ldots, \mathcal{I}_{i-1,n}^k$, respectively ($i = 1, \ldots, s$).

Thus the PDAS algorithm for $(\mathbf{QP}_k)$ reads as following.

Algorithm 7.1.3.

1. **set** $\mathcal{I}_0^k = \Gamma$

2. **for** $i = 1, \ldots, s$

3. **set** $n = 0$, **choose** *initial value* $w_0 = (y_0, u_0, p_0, \mu_{1,0}, \ldots, \mu_{s,0})$ and $c_i > 0$

4. **while** $n < n_{max}$ **do**

   a) **define**
   $$\mathcal{A}_{i,n}^k := \{\xi \in \mathcal{I}_{i-1,n}^k : \mu_{i,n} + c_i\,[g_i(x^k) + g_{i,x}(x^k)(x_n - x^k)] > 0\}$$
   $$\mathcal{I}_{i,n}^k := \mathcal{I}_0^k \setminus \bigcup_{j=1}^{i} \mathcal{A}_{j,n}^k$$

   b) **solve** *the optimality system for* $(\mathbf{QP}_k)$ *and* **find** w_{n+1}

   c) **stop** *or set* $n := n+1$

5. **end while** *and* **fix** $\mathcal{A}_{i,n}^k$ *and* $\mathcal{I}_{i,n}^k$

6. **end for**

7.2 Discrete SQP algorithm

The practical implementation of the SQP algorithm discussed in Chapter 6 requires the discretization strategy. As a general method for numerical solution of differential equations, we consider the *finite element method* (FEM). This method is based on the *Galerkin method* and *piecewise polynomial approximation*. In the Galerkin method, the approximate solution is determined as the element from a certain finite-dimensional vector space of *trial functions* for which the residual error is orthogonal to a certain set of *test functions*.

In FEM, the trial and test functions are piecewise polynomials. A *piecewise polynomial* is a function that is equal to a polynomial on each subdomain of a partition of a given domain. The subdivision is referred to as a *mesh* and the subdomains as *elements*. The basic aspects of FEM can be found in [11, 15, 24].

Let $\overline{\Omega}$ be subdivided into a finite number of subsets T_i (called *elements*). A *triangulation* $\mathcal{T}_h$ denotes a finite non-overlapping subdivision $\mathcal{T}_h = \{T_i\}_{i \in \mathcal{I}}$ of Ω into elements T_i of simple geometry. The *discretization parameter* $h > 0$ is the maximal diameter of all $T_i \in \mathcal{T}_h$. The *set of edges* we denote by $\mathcal{E} = \{E_i\}$. In the following, we restrict ourselves to a triangulation by curvilinear triangles.

Let n_Ω and n_Γ denote the number of the nodes belonging to $\overline{\Omega}$ and lying on the interface Γ, respectively, and let ξ_i be the finite set of all vertices of elements of $\mathcal{T}_h$, which is numbered in a given order. The functions
$$\varphi_{i,\Omega}(\xi_j) = \delta_{ij}, \quad i,j = 1,\ldots,n_\Omega$$
$$\varphi_{i,\Gamma}(\xi_j) = \delta_{ij}, \quad i,j = 1,\ldots,n_\Gamma$$

7 Numerical realization

denote the basis functions (called *nodal basis*) corresponding to the vertices in $\overline{\Omega}$ and on the boundary Γ, respectively. We introduce the finite dimensional spaces

$$V_h := span\{\varphi_{i,\Omega}\}_{i=1}^{n_\Omega}, \quad U_h := span\{\varphi_{i,\Gamma}\}_{i=1}^{n_\Gamma} \quad U_h^* := span\{\varphi_{i,\Gamma}^*\}_{i=1}^{n_\Gamma}.$$

V_h and U_h are the spaces of globally continuous linear functions on each element $T \in \mathcal{T}_h$ and $E \in \mathcal{E}$, respectively, and U_h^* is a space of Dirac measures on $E \in \mathcal{E}$, i.e.,

$$\varphi_{i,\Gamma}^*(\xi_j) = \delta_{\xi_i}(\xi_j) = \delta_{ij} \quad (i,j = 1, \ldots, n_\Gamma).$$

Note V_h is a subspace of $H^1(\Omega) \cap C(\overline{\Omega})$ and U_h is subspace of $L^2(\Gamma)$. We can express y_h, p_h, u_h and $\mu_{i,h}$ corresponding to the unknowns y, p, u and μ_i as linear combinations with respect to these bases. In particular,

$$y_h = \sum_{j=1}^{n_\Omega} y_j \varphi_{j,\Omega}, \quad p_h = \sum_{j=1}^{n_\Omega} p_j \varphi_{j,\Omega}, \quad u_h = \sum_{j=1}^{n_\Gamma} u_j \varphi_{j,\Gamma}, \quad \mu_{i,h} = \sum_{j=1}^{n_\Gamma} \mu_{i,j} \varphi_{j,\Gamma}^*,$$

(7.2.1)

$$(i = 1, \ldots, s)$$

where $\vec{y} = (y_1, \ldots, y_{n_\Omega})^\top$, $\vec{p} = (p_1, \ldots, p_{n_\Omega})^\top$, $\vec{u} = (u_1, \ldots, u_{n_\Gamma})^\top$ and $\vec{\mu}_i = (\mu_{i,1}, \ldots, \mu_{i,n_\Gamma})^\top$ denote the coefficient vectors of y_h, p_h, u_h and $\mu_{i,h}$, respectively. In the similar fashion, we express k-iterates by

$$y_h^k = \sum_{i=1}^{n_\Omega} y_i^k \varphi_{i,\Omega}, \quad p_h^k = \sum_{i=1}^{n_\Omega} p_i^k \varphi_{i,\Omega}, \quad u_h^k = \sum_{i=1}^{n_\Gamma} u_i^k \varphi_{i,\Gamma} \quad \mu_{i,h}^k = \sum_{j=1}^{n_\Gamma} \mu_{i,j}^k \varphi_{j,\Gamma}^*,$$

(7.2.2)

$$(i = 1, \ldots, s)$$

where $\vec{y}^k = (y_1^k, \ldots, y_{n_\Omega}^k)^\top$, $\vec{p}^k = (p_1^k, \ldots, p_{n_\Omega}^k)^\top$, $\vec{u}^k = (u_1^k, \ldots, u_{n_\Gamma}^k)^\top$ and $\vec{\mu}_i^k = (\mu_{i,1}^k, \ldots, \mu_{i,n_\Gamma}^k)^\top$ are the coefficient vectors of y_h^k, p_h^k, u_h^k and $\mu_{i,h}^k$, respectively. As usual we abbreviate

$$\vec{x} = (\vec{y}, \vec{u})^\top, \quad \vec{x}^k = (\vec{y}^k, \vec{u}^k)^\top, \quad x_h = (y_h, u_h), \quad x_h^k = (y_h^k, u_h^k)$$

and

$$\vec{w} = (\vec{y}, \vec{u}, \vec{p}, \vec{\mu}_1, \ldots, \vec{\mu}_s)^\top, \quad \vec{w}^k = (\vec{y}^k, \vec{u}^k, \vec{p}^k, \vec{\mu}_1^k, \ldots, \vec{\mu}_s^k)^\top,$$

$$w_h = (y_h, u_h, p_h, \mu_{1,h}, \ldots, \mu_{s,h}), \quad w_h^k = (y_h^k, u_h^k, p_h^k, \mu_{1,h}^k, \ldots, \mu_{s,h}^k).$$

In order to compute the solution of the linear-quadratic subproblem ($\mathbf{QP}_k$) we use a so-called *"optimize-then-discretize"* approach. That means, we discretize the optimality system for ($\mathbf{QP}_k$), given on the page 56, and compute the

7.2 Discrete SQP algorithm

element $w_h = (y_h, u_h, p_h, \mu_{1,h}, \ldots, \mu_{s,h}) \in V_h \times U_h \times V_h \times [U_h^*]^s$ as a solution of the discretized optimality system (7.2.8) below.

FEM for PDEs is based on their variational forms.

We consider the variational formulation of the state equation (6.1.1), i.e.,

$$a[y_h, v] + (d(y_h^k), v)_\Omega + (b(y_h^k), v)_\Gamma + (d_y(y_h^k)(y_h - y_h^k), v)_\Omega$$
$$+ (b_y(y_h^k)(y_h - y_h^k), v)_\Gamma = (f_h, v)_\Omega + (u_h, v)_\Gamma. \quad (7.2.3)$$

We choose the test function v as an element of V_h. Therefore, thanks to (7.2.1) we obtain

$$\sum_{i=1}^{n_\Omega} y_i \, a[\varphi_{i,\Omega}, \varphi_{j,\Omega}] + (d(\sum_{i=1}^{n_\Omega} y_i^k \varphi_{i,\Omega}), \varphi_{j,\Omega})_\Omega + (b(\sum_{i=1}^{n_\Gamma} y_i^k \varphi_{i,\Gamma}), \varphi_{j,\Omega})_\Gamma$$
$$(d_y(\sum_{i=1}^{n_\Omega} y_i^k \varphi_{i,\Omega}) \sum_{i=1}^{n_\Omega} y_i^k \varphi_{i,\Omega}, \varphi_{j,\Omega})_\Omega + (b_y(\sum_{i=1}^{n_\Gamma} y_i^k \varphi_{i,\Gamma}) \sum_{i=1}^{n_\Gamma} y_i^k \varphi_{i,\Gamma}, \varphi_{j,\Omega})_\Gamma$$
$$= \sum_{i=1}^{n_\Omega} f_i \left(\varphi_{i,\Omega}, \varphi_{j,\Omega} \right)_\Omega + \sum_{i=1}^{n_\Gamma} u_i \left(\varphi_{i,\Gamma}, \varphi_{j,\Omega} \right)_\Gamma, \quad (j = 1, \ldots, n_\Omega)$$

where $\vec{f} = (f_1, \ldots, f_{n_\Omega})^\top$ denotes the coefficient vectors of f_h obtained by projection f into V_h. We restrict our further calculations to the case $a[\varphi_{i,\Omega}, \varphi_{j,\Omega}] = (\nabla \varphi_{i,\Omega}, \nabla \varphi_{j,\Omega})_\Omega$. With the usual abbreviations

$$M := \{(\varphi_{i,\Omega}, \varphi_{j,\Omega})_\Omega\}_{i,j=1}^{n_\Omega \times n_\Omega}, \qquad \text{-- mass matrix}$$
$$K := \{(\nabla \varphi_{i,\Omega}, \nabla \varphi_{j,\Omega})_\Omega\}_{i,j=1}^{n_\Omega \times n_\Omega}, \qquad \text{-- stiffness matrix}$$
$$M^u := \{(\varphi_{i,\Gamma}, \varphi_{j,\Gamma})_\Gamma\}_{i,j=1}^{n_\Gamma \times n_\Gamma},$$
$$M^{yu} := \{(\varphi_{i,\Omega}, \varphi_{j,\Gamma})_\Gamma\}_{i,j=1}^{n_\Omega \times n_\Gamma}, \qquad M^{uy} = (M^{yu})^\top$$

the first term and right hand side in (7.2.3) read as $K\vec{y}$ and $M^{yu}\vec{u} + M\vec{f}$, respectively.

The nonlinear functions d and b make the evaluation of the integrals over Ω and Γ cumbersome. Therefore, these integrals are usually split into sums of integrals over triangles $T \in \mathcal{T}_h$ or edges $E \in \mathcal{E}$, and then each integral is approximated by a *quadrature formula*

$$\int_T f(\xi) d\xi \approx \sum_{k=1}^{K} w_k f(\xi_k) \qquad \int_E f(\xi) d\xi \approx \sum_{k=1}^{\hat{K}} \hat{w}_k f(\hat{\xi}_k) \qquad (7.2.4)$$

with w_k and ξ_k denoting the weights and reference points in the triangle, respectively, and $\hat{w}_k$ and $\hat{\xi}_k$ denoting the respective terms for the edges. The

7 Numerical realization

choice of weight and points determines the formula's order of accuracy. Common quadrature schemes are given in [15, Chapter 4]. The quadrature formula is usually designed to yield the same order of accuracy in terms of mesh size as can be expected of the finite element solutions, [20].

Applying (7.2.4) to the integral of d over Ω we have

$$\int_\Omega d\big(\xi, \sum_{i=1}^{n_\Omega} y_i^k\, \varphi_{i,\Omega}(\xi)\big)\, \varphi_{j,\Omega}(\xi)d\xi \approx \sum_{T\in\mathcal{T}_h}\int_T d\big(\xi, \sum_{i=1}^{n_\Omega} y_i^k\, \varphi_{i,\Omega}(\xi)\big)\, \varphi_{j,\Omega}(\xi)d\xi$$

$$\approx \sum_{T\in\mathcal{T}_h}\sum_{k=1}^{K} w_k\, d\big(\xi_k, \sum_{i=1}^{n_\Omega} y_i^k\, \varphi_{i,\Omega}(\xi_k)\big)\, \varphi_{j,\Omega}(\xi_k). \qquad (j=1,\ldots,n_\Omega) \quad (7.2.5)$$

In a similar fashion we obtain an approximation for the integral of b over Γ

$$\int_\Gamma b\big(\xi, \sum_{i=1}^{n_\Gamma} y_i^k\, \varphi_{i,\Gamma}(\xi)\big)\, \varphi_{j,\Omega}(\xi)d\xi \approx \sum_{E\in\mathcal{E}}\int_E b\big(\xi, \sum_{i=1}^{n_\Gamma} y_i^k\, \varphi_{i,\Gamma}(\xi)\big)\, \varphi_{j,\Omega}(\xi)d\xi$$

$$\approx \sum_{E\in\mathcal{E}}\sum_{k=1}^{\hat{K}} \hat{w}_k\, b\big(\xi_k, \sum_{i=1}^{n_\Gamma} y_i^k\, \varphi_{i,\Omega}(\hat\xi_k)\big)\, \varphi_{j,\Gamma}(\hat\xi_k) \qquad (j=1,\ldots,n_\Omega) \quad (7.2.6)$$

Applying the FE discretization and quadrature formulas (7.2.4) to the terms with derivatives d_y and b_y, we obtain matrices $D_y(\vec{y}^k)$ and $B_y(\vec{y}^k)$, respectively.

Using these approximations for nonlinear terms in (7.2.3) we obtain the finite element discretization of the state equation, which in compact matrix form reads as

$$K\vec{y} + D(\vec{y}^k) + B(\vec{y}^k) + D_y(\vec{y}^k)(\vec{y}-\vec{y}^k) + B_y(\vec{y}^k)(\vec{y}-\vec{y}^k) = M^{yu}\vec{u} + M\vec{f}.$$

$$(7.2.7)$$

Thanks to (4.2.1) the adjoint equation associated with (7.2.3) reads as

$$a[v,p_h] + (d_y(y_h^k)p_h, v)_\Omega + (b_y(y_h^k)p_h, v)_\Gamma + (\phi_y(y_h^k), v)_\Omega + (\psi_y(x_h^k), v)_\Gamma$$
$$+ (\mathcal{L}_{yy}(w_h^k)(y_h-y_h^k), v)_\Gamma + (\mathcal{L}_{yu}(w_h^k)(u_h-u_h^k), v)_\Gamma + \sum_{i=1}^{s}(g_{i,y}(x_h^k)\mu_{i,h}, v)_\Gamma = 0.$$

Setting $v = \varphi_{j,\Omega}$, $j=1,\ldots,n_\Gamma$ we obtain the FE discretization of the last equation, the matrix form of which is

$$\big[K + D_y(\vec{y}^k) + B_y(\vec{y}^k)\big]^\top \vec{p} + \Phi_y(\vec{y}^k) + \Psi_y(\vec{x}^k)$$
$$+ L_{yy}(\vec{w}^k)(\vec{y}-\vec{y}^k) + L_{yu}(\vec{w}^k)(\vec{u}-\vec{u}^k) + \sum_{i=1}^{s} G_{i,y}(\vec{x}^k)\vec{\mu}_i = 0.$$

7.2 Discrete SQP algorithm

Here the matrices $L_{yy}(\vec{y}^k)$, $L_{yu}(\vec{y}^k)$ and $G_{i,y}(\vec{x}^k)$ correspond to the finite element approximation of the terms $\mathcal{L}_{yy}(y^k)$, $\mathcal{L}_{yu}(y^k)$ and $g_{i,y}(x^k)$ obtained analogously to (7.2.5) and (7.2.6). We test the gradient equation

$$\psi_u(x_h^k) + \mathcal{L}_{uy}(w_h^k)(y_h - y_h^k) + \mathcal{L}_{uu}(w_h^k)(u_h - u_h^k) - p_h + \sum_{i=1}^{s} g_{i,u}(x_h^k)\mu_{i,h}$$
$$+ \sum_{i=1}^{s} g_{i,uy}(y^k, u^k)(y - y^k)\mu_i^k + \sum_{i=1}^{s} g_{i,uu}(y^k, u^k)(u - u^k)\mu_i^k = 0, \quad \text{a.e. on } \Gamma$$

by nodal functions $\varphi_{j,\Gamma} \in U_h$

$$(\psi_u(x_h^k), \varphi_{j,\Gamma})_\Gamma + (\mathcal{L}_{uy}(w_h^k)(y_h - y_h^k), \varphi_{j,\Gamma})_\Gamma + (\mathcal{L}_{uu}(w_h^k)(u_h - u_h^k), \varphi_{j,\Gamma})_\Gamma$$
$$- (p_h, \varphi_{j,\Gamma})_\Gamma + \sum_{i=1}^{s}(g_{i,u}(x_h^k)\mu_{i,h}, \varphi_{j,\Gamma})_\Gamma = 0.$$
$$(j = 1, \ldots, n_\Gamma)$$

Owing to quadrature formulas (7.2.4) and the finite element approach we obtain the discretization of the last equation in matrix form

$$\Psi_u(\vec{x}^k) + L_{uy}(\vec{w}^k)(\vec{y} - \vec{y}^k) + L_{uu}(\vec{w}^k)(\vec{u} - \vec{u}^k)$$
$$- M^{uy}\vec{p} + \sum_{i=1}^{s} G_{i,u}(\vec{x}^k)\vec{\mu}_i = 0,$$

where the vector $\Psi_u(\vec{x}^k)$ and matrix $G_{i,u}(\vec{x}^k)$ correspond to the finite element approximations of $\psi_u(x^k)$ and $g_{i,u}(x^k)$ obtained analogously to (7.2.6).

Finally, using the non-smooth representation (7.1.2) we find the discretization of the complementarity conditions (7.1.1) for $(\mathbf{QP}_k)$. Due to the pointwise relation the discrete matrix form of (7.1.2) reads as

$$\vec{\mu}_i - \max\left\{0, \vec{\mu}_i + c_i\left[G_i(\vec{x}^k) + G_{i,y}(\vec{x}^k)(\vec{y} - \vec{y}^k) + G_{i,u}(\vec{x}^k)(\vec{u} - \vec{u}^k)\right]\right\} = 0.$$

7 Numerical realization

Thus, the discretization of the optimality system for $(\mathbf{QP}_k)$ leads to

$$\begin{cases} [K + D_y(\vec{y}^k) + B_y(\vec{y}^k)]^\top \vec{p} + \Phi_y(\vec{y}^k) + \Psi_y(\vec{x}^k) \\ \quad + L_{yy}(\vec{w}^k)(\vec{y} - \vec{y}^k) + L_{yu}(\vec{w}^k)(\vec{u} - \vec{u}^k) + \sum_{i=1}^{s} G_{i,y}(\vec{x}^k)\vec{\mu}_i = 0 \\ \Psi_u(\vec{x}^k) + L_{uy}(\vec{w}^k)(\vec{y} - \vec{y}^k) + L_{uu}(\vec{w}^k)(\vec{u} - \vec{u}^k) \\ \quad - M^{uy}\vec{p} + \sum_{i=1}^{s} G_{i,u}(\vec{x}^k)\vec{\mu}_i = 0 \\ [K + D_y(\vec{y}^k) + B_y(\vec{y}^k)]\vec{y} - M^{yu}\vec{u} \\ \quad + M\vec{f} + D(\vec{y}^k) + B(\vec{y}^k) - D_y(\vec{y}^k)\vec{y}^k - B_y(\vec{y}^k)\vec{y}^k = 0 \\ \vec{\mu}_i - \max\left\{0, \vec{\mu}_i + c_i\left[G_i(\vec{x}^k) + G_{i,y}(\vec{x}^k)(\vec{y} - \vec{y}^k) + G_{i,u}(\vec{x}^k)(\vec{u} - \vec{u}^k)\right]\right\} = 0. \end{cases}$$

(7.2.8)

and the PDAS method applied to $(\mathbf{QP}_k)$ reads now as [1]

Algorithm 7.2.1.

1. set $\mathcal{I}_0^k = \{1, \ldots, n_\Gamma\}$

2. **for** $i = 1, \ldots, s$

3. set $n = 0$, choose $\vec{w}_0 = (\vec{y}_0, \vec{u}_0, \vec{p}_0, \vec{\mu}_{1,0}, \ldots, \vec{\mu}_{s,0})$ and $c_i > 0$

4. **while** $n < n_{max}$ **do**

   a) **define**

   $$\mathcal{A}_{i,n}^k := \{j \in \mathcal{I}_{i-1,n}^k : [\vec{\mu}_{i,n} + c_i\left(G_i(\vec{x}^k) + G_{i,x}(\vec{x}^k)(\vec{x}_n - \vec{x}^k)\right)]_j > 0\},$$

   $$\mathcal{I}_{i,n}^k := \mathcal{I}_0^k \setminus \bigcup_{j=1}^{i} \mathcal{A}_{j,n}^k$$

   b) **solve** *the optimality system (7.2.8) and find* $\vec{w}_{n+1}$

   c) **stop** [2] *or set* $n := n + 1$

5. **end while** *and fix* $\mathcal{A}_{i,n}^k$ *and* $\mathcal{I}_{i,n}^k$

6. **end for**

---
[1] compare with Algorithm 7.1.3
[2] The stopping criterion is $\mathcal{A}_{i,n}^k = \mathcal{A}_{i,n+1}^k$, see [9].

7.2 Discrete SQP algorithm

Note the optimality system (7.2.8) is equivalent to the equation (7.2.9) below.

$$\begin{pmatrix} L_{yy}(\vec{w}^k) & L_{yu}(\vec{w}^k) & [K+D_y(\vec{y}^k)+B_y(\vec{y}^k)]^T & G_{1,y}(\vec{x}^k) & \cdots & G_{s,y}(\vec{x}^k) \\ L_{uy}(\vec{w}^k) & L_{uu}(\vec{w}^k) & -M^{uy} & G_{1,u}(\vec{x}^k) & \cdots & G_{s,u}(\vec{x}^k) \\ K+D_y(\vec{y}^k)+B_y(\vec{y}^k) & -M^{yu} & 0 & 0 & \cdots & 0 \\ -c_1 I_{\mathcal{A}_{1,n}^k} G_{1,y}(\vec{x}^k) & -c_1 I_{\mathcal{A}_{1,n}^k} G_{1,u}(\vec{x}^k) & 0 & I_{\mathcal{I}_{1,n}^k} & \cdots & 0 \\ \cdots & \cdots & \cdots & \cdots & \cdots & \cdots \\ -c_s I_{\mathcal{A}_{s,n}^k} G_{s,y}(\vec{x}^k) & -c_s I_{\mathcal{A}_{s,n}^k} G_{s,u}(\vec{x}^k) & 0 & 0 & \cdots & I_{\mathcal{I}_{s,n}^k} \end{pmatrix} \begin{pmatrix} \vec{y} \\ \vec{u} \\ \vec{p} \\ \vec{\mu}_1 \\ \cdots \\ \vec{\mu}_s \end{pmatrix}$$

$$= - \begin{pmatrix} \Phi_y(\vec{y}^k)+\Psi_y(\vec{x}^k)-L_{yy}(\vec{w}^k)\vec{y}^k-L_{yu}(\vec{w}^k)\vec{u}^k \\ \Psi_u(\vec{x}^k)-L_{uy}(\vec{w}^k)\vec{y}^k-L_{uu}(\vec{w}^k)\vec{u}_k \\ D(\vec{y}^k)+B(\vec{y}^k)-M\vec{f}-D_y(\vec{y}^k)\vec{y}^k-B_y(\vec{y}^k)\vec{y}^k \\ -G_1(\vec{x}^k)+c_1 I_{\mathcal{A}_{1,n}^k} G_{1,y}(\vec{x}^k)\vec{y}^k+c_1 I_{\mathcal{A}_{1,n}^k} G_{1,u}(\vec{x}^k)\vec{u}^k \\ \cdots \\ -G_s(\vec{x}^k)+c_s I_{\mathcal{A}_{s,n}^k} G_{s,y}(\vec{x}^k)\vec{y}^k+c_s I_{\mathcal{A}_{s,n}^k} G_{s,u}(\vec{x}^k)\vec{u}^k \end{pmatrix} \quad (7.2.9)$$

The matrices $I_{\mathcal{I}_{i,n}^k}$ and $I_{\mathcal{A}_{i,n}^k}$, $i=1,\ldots,s$, are diagonal matrices with $\{0,1\}$ entries, and they are related to the indices where the max-functions are "active" or "inactive" (equal zero) associated with the current iterate $\vec{w}^k$, i.e.,

$$I_{\mathcal{A}_{i,n}^k} := diag(\chi_{\mathcal{A}_{i,n}^k}),$$
$$I_{\mathcal{I}_{i,n}^k} := diag(\chi_{\mathcal{I}_i^k}) = I - I_{\mathcal{A}_{i,n}^k}.$$

Summarizing all preparatory notions, the SQP algorithm can be presented by

Algorithm 7.2.2.

1. **choose** *the starting point* $\vec{w}^0$

2. **for** $k=0,\ldots,k_{max}$

3. **solve** $(\mathbf{QP}_k)$ *and* **find** $\vec{w}^{k+1}$ *using Algorithm 7.2.1*

4. **check** *stopping criterion for SQP*

5. **set** $k=k+1$

6. **end for**

7 Numerical realization

The residual of the optimality system for (**P**) serves as a convergence criterion for the SQP method, i.e., if the norm of

$$\begin{pmatrix} [K + D_y(\vec{y}^{k+1})]^\top \vec{p}^{k+1} + \Phi_y(\vec{y}^{k+1}) + \Psi_y(\vec{x}^{k+1}) + \sum_{i=1}^{s} G_{i,y}(\vec{x}^{k+1})\vec{\mu}^{k+1} \\ \Psi_u(\vec{x}^{k+1}) - M^{uy}\vec{p}^{k+1} + \sum_{i=1}^{s} G_{i,u}(\vec{x}^{k+1})\vec{\mu}^{k+1} \\ K\vec{y}^{k+1} + D(\vec{y}^{k+1}) - M^{yu}\vec{u}^{k+1} \\ \vec{\mu}_1^{k+1} - \max\left\{0, \vec{\mu}_1^{k+1} + c_1\, G_1(\vec{x}^{k+1})\right\} \\ \cdots \\ \vec{\mu}_s^{k+1} - \max\left\{0, \vec{\mu}_s^{k+1} + c_s\, G_s(\vec{x}^{k+1})\right\} \end{pmatrix}$$

is small enough[1] then $\vec{w}^{k+1}$ is the discretized optimal solution.

7.3 Numerical tests

The SQP method presented in the previous section has been numerically tested on a set of various optimal control problems. It was observed that the sequences of the state and control functions converge quadratically. Below, we show these results on the paradigm of model problems in a unit disk Ω.

7.3.1 An example of boundary optimal control problems

Example 7.3.1. *We consider a boundary optimal control problem governed by nonlinear mixed constraints. According to the problem setting* (**P**), *the task is to determine a control* $u \in L^\infty(\Gamma)$ *that minimizes the objective*

$$\frac{1}{2}\|y - y_{d,\Omega}\|^2_{L^2(\Omega)} + \frac{1}{2}\|y - y_{d,\Gamma}\|^2_{L^2(\Gamma)} + \frac{1}{2}\|u - u_d\|^2_{L^2(\Gamma)}$$

subject to the elliptic state equation with Neumann boundary condition

$$-\Delta y + y + y^3 = f \quad \text{in } \Omega,$$
$$\partial_n y = u + e_\Gamma \quad \text{on } \Gamma,$$

and nonlinear in state mixed control-state constraints

$$g_1(y, u) = u + y + y^3 - y_c \leqslant 0$$
$$g_2(y, u) = u \leqslant 0 \quad \text{a.e. on } \Gamma.$$

---
[1] One requires $\|residual\| < \rho$, where the choice of ρ agrees with the discretization error.

7.3 Numerical tests

We use the following data

$$f = -\Delta y + y + y^3 = -6(r-1) - \frac{3(r-1)^2}{r} + (r-1)^3 + (r-1)^9,$$

$$y_{d,\Omega} = -\Delta p + p + 3y^2 p + y = -6(r-1) - \frac{3(r-1)^2}{r} + 3(r-1)^9 + 2(r-1)^3,$$

$$y_c = \begin{cases} -1, & \text{if } \varphi \in [0, \frac{\pi}{2}], \\ \cos(2\varphi), & \text{if } \varphi \in [\frac{\pi}{2}, \pi], \\ 1, & \text{if } \varphi \in [\pi, \frac{3\pi}{2}], \\ -\cos(2\varphi), & \text{if } \varphi \in [\frac{3\pi}{2}, 2\pi], \end{cases} \qquad u_d = \begin{cases} -1 + \sin(2\varphi), & \text{if } \varphi \in [0, \frac{\pi}{2}], \\ -\sin(\varphi), & \text{if } \varphi \in [\frac{\pi}{2}, \pi], \\ \sin(2\varphi), & \text{if } \varphi \in [\pi, \frac{3\pi}{2}], \\ -\cos(\varphi), & \text{if } \varphi \in [\frac{3\pi}{2}, 2\pi], \end{cases}$$

$$e_\Gamma = \begin{cases} 1, & \text{if } \varphi \in [0, \frac{\pi}{2}], \\ \sin(\varphi), & \text{if } \varphi \in [\frac{\pi}{2}, \pi], \\ 0, & \text{if } \varphi \in [\pi, \frac{3\pi}{2}], \\ \cos(\varphi), & \text{if } \varphi \in [\frac{3\pi}{2}, 2\pi], \end{cases} \qquad y_{d,\Gamma} = \begin{cases} \sin(2\varphi), & \text{if } \varphi \in [0, \frac{\pi}{2}], \\ 0, & \text{else.} \end{cases}$$

It is straightforward to verify that

$$y^* = (r-1)^3, \quad u^* = -e_\Gamma, \quad p^* = y^*,$$

$$\mu_1^* = y_{d,\Gamma}, \qquad \mu_2^* = \begin{cases} \sin(2\varphi), & \text{if } \varphi \in [\pi, \frac{3\pi}{2}], \\ 0, & \text{else.} \end{cases}$$

satisfy the optimality conditions (4.1.6a)–(4.1.6d). Moreover, Assumptions **(A1)**–**(A3)** are easily confirmed. To show the linearized Slater condition, we observe that the linearized equation

$$-\Delta \hat{y} + (1 + 3(y^*)^2)\,\hat{y} = 0 \quad \text{in } \Omega,$$
$$\partial_n \hat{y} = \hat{u} \quad \text{in } \Gamma$$

satisfies the maximum principle, i.e., $\hat{u} \leqslant 0$ on Γ implies $\hat{y} \leqslant 0$ in Ω. Hence, choosing $\hat{u} \equiv -1$ yields

$$g_1(y^*, u^*) + \hat{u} + (1 + 3(y^*)^2)\hat{y} \leqslant -1,$$
$$g_2(y^*, u^*) + \hat{u} \leqslant -1,$$

and **(A4)** holds.

The security sets S_1 and S_2 do not intersect for values $\sigma_1, \sigma_2 \in (0, \frac{1}{\sqrt{2}})$. More-

7 Numerical realization

over, (4.1.7) is uniquely solvable since

$$g_{1,u}^{-1}(y^*, u^*)g_{1,y}(y^*, u^*) = 1 + 3(y^*)^2,$$
$$g_{2,u}^{-1}(y^*, u^*)g_{2,y}(y^*, u^*) = 0,$$

i.e., **(A6)** holds as well. Finally, we verify the second-order sufficient condition **(A7)**

$$\mathcal{L}_{xx}(w^*)(\delta x, \delta x) = \int_\Omega (1 + 6y^*p^*)(\delta y)^2 d\xi + \int_\Gamma (1 + 6y^*\mu_1^*)(\delta y)^2 ds + \int_\Gamma (\delta u)^2 ds$$
$$\geq \|\delta y\|_{L^2(\Omega)}^2 + \|\delta u\|_{L^2(\Gamma)}^2.$$

We discretized the problem using linear finite elements on a triangular mesh with 2097 nodes and 4064 triangles. We used $y^0 = -1.2$, $u^0 \equiv -1$ and $p^0 = \mu_1^0 = \mu_2^0 \equiv 0$ as an initial guess. The inequality constrained subproblems **(QP$_k$)** were solved by using the PDAS strategy. Due to Theorem 6.3.7, we expect quadratic convergence of the SQP method.

Table 7.3.1 below shows the convergence behavior of the discretized method. In the last column we used the norm $H^1(\Omega) \times L^\infty(\Gamma) \times H^1(\Omega) \times [L^\infty(\Gamma)]^2$ and observe quadratic convergence of the considered method with constant bounded by ≈ 0.07. Figures 7.3.1–7.3.3 display the solution obtained.

k	$\|y^5 - y^k\|_{H^1(\Omega)}$	$\|u^5 - u^k\|_{L^\infty(\Gamma)}$	$\frac{\|w^5 - w^k\|}{\|w^5 - w^{k-1}\|^2}$
0	2.36e+0	6.73e+0	—
1	9.34e-1	3.68e+0	7.01e-2
2	2.22e-1	1.54e+0	7.05e-2
3	3.73e-3	2.17e-2	2.91e-3
4	5.10e-7	2.27e-6	1.56e-3

Table 7.3.1: The SQP iterates for semilinear boundary optimal control problem subject to nonlinear mixed constraints.

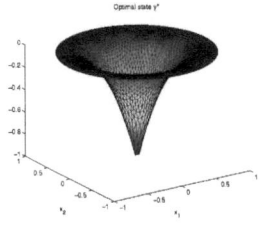

Figure 7.3.1: State y

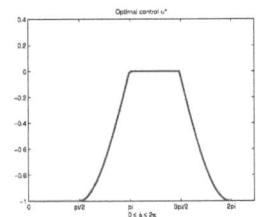

Figure 7.3.2: Control u

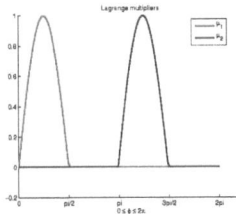

Figure 7.3.3: Multipliers μ_1 and μ_2

7.3.2 An example of distributed optimal control problems

Example 7.3.2. Let Ω be a unit ball. We consider a distributed control problem governed by a semilinear elliptic equation with homogeneous Dirichlet boundary condition and linear constraints

$$\text{Minimize} \quad \frac{1}{2}\|y - y_d\|_{L^2(\Omega)}^2 + \frac{1}{2}\|u - u_d\|_{L^2(\Omega)}^2$$

$$\text{subject to} \quad \begin{cases} -\Delta y + y + y^3 = u + e_\Omega & \text{in } \Omega, \\ y = 0 & \text{on } \Gamma \end{cases}$$

$$\text{and} \quad \begin{cases} g_1(y,u) = \varepsilon u + y - y_c \leqslant 0, \\ g_2(y,u) = u \leqslant 0 \end{cases} \quad a.e. \text{ on } \Omega.$$

Note this model fits to the problem setting (**P**′). We set

$$e_\Omega := -\max\{-1, \min\{0, 3r - 2\}\},$$

7 Numerical realization

$$y_d := \begin{cases} 1 - 3r, & r > \frac{1}{3}, \\ 0, & r \leqslant \frac{1}{3}, \end{cases} \qquad u_d := \begin{cases} 3r - 2, & r > \frac{1}{3}, \\ -1 + \varepsilon(1 - 3r), & r \leqslant \frac{1}{3}, \end{cases}$$

$$y_c := \begin{cases} -\varepsilon, & r < \frac{1}{3}, \\ 3r - 1 + \varepsilon(3r - 2), & r \in [\frac{1}{3}, \frac{2}{3}], \\ 1, & r > \frac{2}{3}. \end{cases}$$

where $r^2 = x_1^2 + x_2^2$. Note, $|r| \leqslant 1$ for all x_1, x_2 on Γ.
It is easy to check that

$$y^* \equiv 0, \quad p^* \equiv 0, \qquad u^* = -e_\Omega,$$

$$\mu_1^* := \begin{cases} 1 - 3r, & r < \frac{1}{3}, \\ 0, & r \geqslant \frac{1}{3}, \end{cases} \qquad \mu_2^* := \begin{cases} 3r - 2, & r > \frac{2}{3}, \\ 0, & r \leqslant \frac{2}{3} \end{cases}$$

satisfy the optimality system

$$\begin{cases} \begin{cases} -\Delta y + y + y^3 = u + e_\Omega & \text{in } \Omega, \\ y = 0 & \text{on } \Gamma, \end{cases} \\ u - u_d - p + \varepsilon \mu_1 + \mu_2 = 0 \quad \text{a.e. in } \Omega, \\ \begin{cases} -\Delta p + p + 3y^2 p = -(y - y_d) - \mu_1 & \text{in } \Omega, \\ p = 0 & \text{on } \Gamma, \end{cases} \\ 0 \leqslant \mu_1 \perp \varepsilon u + y - y_c \leqslant 0 \\ 0 \leqslant \mu_2 \perp u \leqslant 0 \end{cases} \quad \text{a.e. in } \Omega.$$

Moreover, Assumptions (**A1'**)–(**A3'**) and (**A5'**) are obviously fulfilled.
To show the linearized Slater condition, we observe that the linearized equation

$$-\Delta \hat{y} + (1 + 3(y^*)^2)\, \hat{y} = \hat{u} \quad \text{in } \Omega,$$
$$\hat{y} = 0 \quad \text{in } \Gamma$$

satisfies the maximum principle, i.e., $\hat{u} \leqslant 0$ on Ω implies $\hat{y} \leqslant 0$ in Ω. Hence, choosing $\hat{u} \equiv -1$ yields

$$g_1(y^*, u^*) + \varepsilon \hat{u} + \hat{y} \leqslant -\varepsilon, \quad (\varepsilon > 0)$$
$$g_2(y^*, u^*) + \hat{u} \leqslant -1,$$

and (**A4'**) holds with $\tau = \min\{1, \varepsilon\}$.

The security sets S_1' and S_2' do not intersect for values $\sigma_1, \sigma_2 \in (0, \frac{1}{2})$, i.e., (**A6'**) holds as well. Finally, we verify the second-order sufficient condition

(**A7'**)

$$\mathcal{L}_{xx}(w^*)(\delta x, \delta x) = \int_\Omega (1 + 6y^* p^*)(\delta y)^2 d\xi + \int_\Omega (\delta u)^2 ds$$
$$= \|\delta y\|^2_{L^2(\Omega)} + \|\delta u\|^2_{L^2(\Omega)}.$$

We discretized the problem using linear finite elements on a triangular mesh with 2097 nodes and 4064 triangles. We used $y^0 = -1.2$, $u^0 \equiv -0.5$ and $p^0 = \mu_1^0 = \mu_2^0 \equiv 0$ as an initial guess. The inequality constrained quadratic subproblems were solved by using the PDAS strategy. Due to Theorem 6.4.3 we expect quadratic convergence of the SQP method.

We computed the discrete solutions for $\varepsilon = 1$. Table 7.3.2 below confirms the good convergence of the discretized method. In the last column we used the norm $H^1(\Omega) \times L^\infty(\Omega) \times H^1(\Omega) \times [L^\infty(\Omega)]^2$. We observe quadratic convergence of the considered method with constant bounded by ≈ 0.049.

The obtained solution is displayed in the figures 7.3.4–7.3.7.

k	$\|y^5 - y^k\|_{H^1(\Omega)}$	$\|u^5 - u^k\|_{L^\infty(\Omega)}$	$\frac{\|w^5-w^k\|}{\|w^5-w^{k-1}\|^2}$
0	2.76e+1	7.79e-1	—
1	4.25e+0	2.88e-1	4.89e-2
2	7.87e-1	1.60e-1	4.80e-2
3	1.35e-2	3.68e-3	1.54e-2
4	4.80e-7	6.22e-7	5.62e-3

Table 7.3.2: The SQP iterates for nonlinear distributed optimal control problem subject to linear constraints.

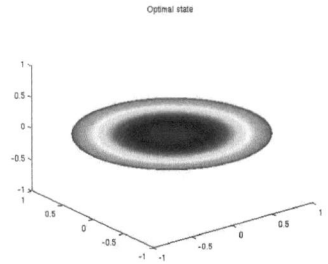

Figure 7.3.4: State y

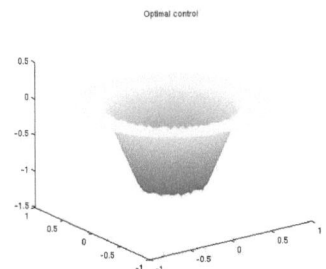

Figure 7.3.5: Control u

7 Numerical realization

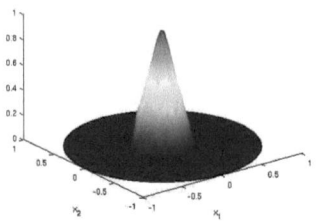

Figure 7.3.6: Multiplier μ_1

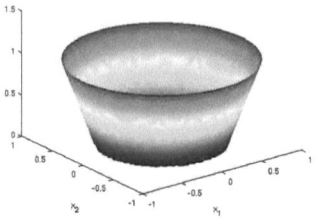

Figure 7.3.7: Multiplier μ_2

8 Additional Material

This chapter recalls some basic results from functional analysis and analysis of partial differential equations, as well as tools that play an essential role in the optimization theory of convex and nonconvex optimal control problems. The proofs are mostly omitted. For a deeper study of these topics we refer to the textbooks of functional analysis [26, 28, 50] and study of Lebesgue and Sobolev spaces [1, 23], of analysis of PDEs [19, 35] and optimization theory [4, 10, 46]. We assume that the reader is already familiar with the basic notions of linear functional analysis. In particular, we make free use of such fundamental concepts as Banach and Hilbert spaces, dual spaces, coercivity and compactness, strong and weak convergence, linear continuous (bounded) operators and their dual operators and so on.

First we recall some elementary inequalities, which are used frequently in the text.

Maximum inequality

$$\max\{a,b\} - \max\{c,d\} \leqslant \max\{a-b, c-d\} \qquad (8.0.1)$$

Proof. For any $a, b, c, d \in \mathbb{R}$ we consider

$$a = a - c + c \leqslant a - c + \max\{c,d\} \leqslant \max\{a-c, b-d\} + \max\{c,d\}$$

Analogous arguments yield $b \leqslant \max\{a-c, b-d\} + \max\{c,d\}$. Finally

$$\max\{a,b\} \leqslant \max\{a-c, b-d\} + \max\{c,d\}$$

holds. □

Young's inequality

Let $1 < p, p' < \infty$, $\frac{1}{p} + \frac{1}{p'} = 1$. Then

$$ab \leqslant \frac{a^p}{p} + \frac{b^{p'}}{p'} \qquad (a, b \geqslant 0). \qquad (8.0.2)$$

8 Additional Material

Proof. For $a = 0$ or $b = 0$, (8.0.2) is clear. The mapping $x \to e^x$ is convex, and consequently for $a, b > 0$

$$ab = e^{\log a + \log b} = e^{\frac{1}{p} \log a^p + \frac{1}{p'} \log b^{p'}} \leqslant \frac{1}{p} e^{\log a^p} + \frac{1}{p'} e^{\log b^{p'}} = \frac{a^p}{p} + \frac{b^{p'}}{p'}.$$

$\square$

Young's inequality with γ

$$ab \leqslant \gamma a^p + C(\gamma) b^{p'} \qquad (a, b > 0, \gamma > 0) \qquad (8.0.3)$$

for $C(\gamma) = (\gamma p)^{-p'/p} {p'}^{-1}$.

Proof. Write $ab = (\gamma p)^{1/p} a \frac{b}{(\gamma p)^{1/p}}$ and apply Young's inequality. $\square$

Hölder's inequality

Assume $1 < p, p' \leqslant \infty$, $\frac{1}{p} + \frac{1}{p'} = 1$. Then if $u \in L^p(\Omega), v \in L^{p'}(\Omega)$, we have

$$\int_\Omega |uv| dx \leqslant \|u\|_{L^p(\Omega)} \|v\|_{L^{p'}(\Omega)}. \qquad (8.0.4)$$

Proof. By homogeneity, we may assume $\|u\|_{L^p} = \|v\|_{L^{p'}} = 1$. Then Young's inequality implies for $1 < p, p' < \infty$ that

$$\int_\Omega |uv| dx \leqslant \frac{1}{p} \int_\Omega |u|^p dx + \frac{1}{p'} \int_\Omega |v|^{p'} dx = 1 = \|u\|_{L^p} \|v\|_{L^{p'}}.$$

$\square$

8.1 Convexity

One of the important role in optimization theory plays convex problems, since the necessary optimality conditions are also sufficient for them. This section gives a short overview of the results from convex analysis and theory of monotone operators.

For the purposes of this section, let X be a vector space and let $C \subset X$ be a convex set according to Definition 8.1.1 below.

Definition 8.1.1 (Convex set). *A set $C \subset X$ is convex if for any two x, y in C, the line segment $\lambda x + (1 - \lambda) y$ is contained in C where $\lambda \in [0, 1]$.*

Definition 8.1.2 (Convex function). *A function $f; C \to \mathbb{R}$ is <u>convex</u> over C if for any $\lambda \in [0,1]$ and any $x, y \in C$*

$$f(\lambda x + (1-\lambda)y) \leqslant \lambda f(x) + (1-\lambda)f(y).$$

A map f is called <u>strictly convex</u> if the inequality above is strict whenever $x \neq y$ and $\lambda \in (0,1)$.

Lemma 8.1.3 (Sum of convex functions). *Let f and g be convex maps over the convex set C. Then their sum is a convex map over C. Moreover, the sum of convex maps is strictly convex whenever one of the maps is strictly convex.*

Lemma 8.1.4 (Weak lower semicontinuity). *Let C be a non-empty, convex and closed subset of a normed space X and $f : C \to \mathbb{R}$ be continuous and convex. Then f is weakly lower semicontinuous , i.e., if $x_n \rightharpoonup x$ in C implies that $\liminf_{n \to \infty} f(x_n) \geqslant f(x)$.*

Lemma 8.1.5 (Minima of convex problem). *Let X be a Hilbert space and $C \subset X$ be a non-empty, convex, closed and bounded subset and f be a lower semicontinuous and convex functional. Then f attains its global minimum over C, i.e., there exists $x_0 \in C$ such that $f(x_0) = \inf_{x \in C} f(x)$. Moreover, the minimizer x_0 is unique if f is strict convex.*

8.2 Bilinear forms

In this section we summarize some elementary properties of bilinear forms on Banach spaces. Bilinear forms will be used for studying **SSC** (see Section 4.2) and elliptic PDEs (see Section 8.7).

Definition 8.2.1. *Let X be a Banach space. The map $a : X \times X \to \mathbb{R}$ is called a <u>bilinear form</u> if, and only if,*

$$a[\lambda_1 x_1 + \lambda_2 x_2, y] = \lambda_1 a[x_1, y] + \lambda_2 a[x_2, y],$$
$$a[x, \lambda_1 y_1 + \lambda_2 y_2] = \lambda_1 a[x, y_1] + \lambda_2 a[x, y_2],$$

hold for each x, x_1, $x_2 \in X$, y, y_1, $y_2 \in X$ and $\lambda_1, \lambda_2 \in \mathbb{R}$.

Definition 8.2.2. *Let $a : X \times X \to \mathbb{R}$ be a bilinear form on X.*

- *a is called <u>bounded</u> (or <u>continuous</u>) if there exists $\bar{c} > 0$ such that*

$$a[y, x] \leqslant \bar{c} \|y\|_X \|x\|_X \quad \text{for all } x, y \in X,$$

8 Additional Material

- *a* is called <u>symmetric</u> if

$$a[y, x] = a[x, y] \quad \text{for all } x, y \in X,$$

- *a* is called <u>coercive</u> (or <u>elliptic</u>) if there exists $\underline{c} > 0$ such that

$$a[x, x] \geq \underline{c} \|x\|_X^2 \quad \text{for all } x \in X,$$

- *a* is called <u>positive semidefinite</u> if

$$a[x, x] \geq 0 \quad \text{for each } x \in X.$$

Next Lemma states the stability property for the bilinear form, for more details see [31].

Lemma 8.2.3. *Let a be a continuous symmetric bilinear form on $X \times X$, H be a subset of X and $\delta > 0$ with*

$$a[h, h] \geq \delta \|h\|^2 \quad \text{for all } h \in H.$$

Then there are $\delta_0 > 0$ and $\gamma > 0$ such that

$$a[h+z, h+z] \geq \delta_0 \|h+z\|^2 \quad \text{for all } h \in H, \ z \in X \text{ and } \|z\| \leq \gamma \|h\|.$$

Proof. Put $b = \|a\|$, i.e., $|a[x, z]| \leq b \|x\| \|z\|$ for all $x, z \in X$, and choose $\gamma > 0$ small enough such that $\delta' = \delta - 2b\gamma - b\gamma^2 > 0$. Then for all $h \in H$ and $z \in X$ with $\|z\| \leq \gamma \|h\|$ one has

$$a[h+z, h+z] \geq \delta \|h\|^2 - 2b \|h\| \|z\| - b \|z\|^2 \geq \delta' \|h\|^2.$$

Because of $\|h+z\| \leq \|h\| + \|z\| \leq (1+\gamma) \|h\|$ we get with $\delta_0 = \delta'/(1+\gamma)^2 > 0$

$$a[h+z, h+z] \geq \delta' \|h\|^2 \geq \delta_0 \|h+z\|^2.$$

$\square$

8.3 Elements of measure theory

Definition 8.3.1. *Let X be an abstract set, and let $\mathcal{M}$ be a family of subsets of X closed under the formation of finite unions and of complements. Let $\overline{\mathcal{M}}$ be the smallest family of sets containing $\mathcal{M}$ and closed under the formation of countable unions and complements.*

Definition 8.3.2. *Let μ be a single-valued function defined on $\mathcal{M}$ such that*

(i) $-\infty < \mu(A) < +\infty$ for all $A \in \mathcal{M}$,

(ii) $\mu(\emptyset) = 0$,

(iii) $\sup\limits_{A \in \mathcal{M}} |\mu(A)| < +\infty$,

(iv) $\mu(A \cup B) = \mu(A) + \mu(B)$ for all $A, B \in \mathcal{M}$ such that $A \cap B = \emptyset$.

Then μ is called <u>finitely additive measure on</u> $\mathcal{M}$. The set of all such measures for fixed X and $\mathcal{M}$ is denoted by $\mathcal{F}(X, \mathcal{M})$.

Definition 8.3.3. *Let μ be an element of $\mathcal{F}$ such that for every sequence $\{A_n\}_{n=1}^{\infty} \subset \mathcal{M}$ such that $A_1 \supset A_2 \supset \cdots \supset A_n \supset \cdots$ and $\bigcap\limits_{n=1}^{\infty} A_n = \emptyset$, we have $\lim_{n \to \infty} \mu(A_n) = 0$. Then μ is called <u>countable additive</u>.*

Definition 8.3.4. *Let μ be a measure in $\mathcal{F}$ such that $\mu \geqslant 0$. If every countably additive measure ν such that $0 \leqslant \nu \leqslant \mu$ is identical zero, then μ is called <u>purely finite additive</u>.*

Now we recall a number of properties possessed by countable additive and purely finite additive measures, for a deeper study we refer to [49].

Theorem 8.3.5. *Let $\mathcal{M} = \overline{\mathcal{M}}$. Then if $\lambda \geqslant 0$ is purely finite additive and $\nu \geqslant 0$ is countably additive, there exists a decreasing sequence $B_1, B_2, \ldots, B_n, \ldots$ of elements of $\mathcal{M}$ such that $\lim_{n \to \infty} \nu(B_n) = 0$ and $\lambda(B_n) = \lambda(\mathcal{M})$ $(n = 1, 2, \ldots)$. Conversely, if $\mu \in \mathcal{M}$ and the above conditions hold for all countably additive ν, then μ is purely finite additive.*

Theorem 8.3.6. *Let μ be a non-negative measure in $\mathcal{M}$. Then there exist a countably additive measure $\mu_c \geqslant 0$ and a purely finite additive measure $\mu_p \geqslant 0$ such that $\mu = \mu_c + \mu_p$.*

Theorem 8.3.7. *Let μ be any measure in $\mathcal{M}$. Then μ can be uniquely written as the sum of a countable additive measure μ_c and a purely finite additive measure μ_p.*

8.4 Function spaces

Assume that $\Omega \subset \mathbb{R}^N$ ($N \in \mathbb{N}$) is some given measurable and bounded set.

8 Additional Material

<u>Lebesgue spaces</u> $L^p(\Omega)$. The space $L^p(\Omega)$ is the *set of all measurable functions* having finite norm

$$\|f\|_{L^p(\Omega)} = \Big(\int_\Omega |f(x)|^p dx\Big)^{\frac{1}{p}} \text{ for all } p \in [1,\infty).$$

For $p = \infty$ we have a Banach space $L^\infty(\Omega)$ of essentially bounded measurable functions with norm defined by

$$\|f\|_{L^\infty(\Omega)} = \operatorname*{ess\,sup}_{x\in\Omega} |f(x)| := \inf_{\mu(F)=0} \sup_{x\in E\setminus F} |f(x)|.$$

We recall that the dual space of $L^p(\Omega)$ for $1 < p < \infty$ is identified with $L^{p'}(\Omega)$, where $\frac{1}{p} + \frac{1}{p'} = 1$. The set of functions measurable on each compact subset of Ω is denoted by $L^1_{loc}(\Omega)$.

If Ω is open, we denote by $C^k(\Omega)$ the set of all functions $f : \Omega \to \mathbb{R}$ such that all partial derivatives up to order $k \in \mathbb{N}$

$$D^\alpha f = \frac{\partial^{\alpha_1 + \ldots + \alpha_N} f}{\partial x_1^{\alpha_1} \cdots \partial x_N^{\alpha_N}}$$

are continuous in Ω, where $\alpha = (\alpha_1, \ldots, \alpha_N) \in \mathbb{N}_0^N$ denotes any multi-index such that its length $|\alpha|$ satisfies $|\alpha| = \alpha_1 + \ldots + \alpha_N \leq k$.

We denote by $C^\infty(\Omega)$ the set of all functions belonging to $C^k(\Omega)$ for any $k \in \mathbb{N}$. The functions in $C^k(\Omega)$ having compact support in Ω form a subspace, which is denoted by $C^k_0(\Omega)$. The space of infinite differentiable functions with compact support in Ω, is denoted by $C^\infty_0(\Omega)$ or $\mathcal{D}(\Omega)$. Its dual space is the space $\mathcal{D}'(\Omega)$ of distributions on Ω, [48].

The space $C^k(\overline{\Omega})$ consists of all those elements of $C^k(\Omega)$ that together with all their derivatives up to order k can be continuously extended onto $\overline{\Omega}$. The space of continuous function on $\overline{\Omega}$ is denoted by $C(\overline{\Omega})$

<u>Sobolev spaces</u> $W^{s,p}(\Omega)$ (see Definition below) are defined over an arbitrary domain $\Omega \subset \mathbb{R}^N$ and are vector subspaces of various Lebesgue spaces $L^p(\Omega)$. The definition of the Sobolev spaces is based on so-called *weak derivatives*.

Definition 8.4.1 (Weak derivative). *Let* $y \in L^1_{loc}(\Omega)$ *and* α *be a multi-index. Suppose there exists a function* $w \in L^1_{loc}(\Omega)$ *such that*

$$\int_\Omega y(x) D^\alpha v(x) dx = (-1)^{|\alpha|} \int_\Omega w(x) v(x) dx$$

holds for all $v \in C^\infty_0(\Omega)$. *The function w is called a <u>weak derivative of y of order α</u> and denoted by* $D^\alpha y$.

Let s be a non-negative integer and let $1 \leq p \leq \infty$. Sobolev space $W^{s,p}(\Omega)$ is the set of all distributions $u \in L^p(\Omega)$ such that $D^\alpha u \in L^p(\Omega)$ for $|\alpha| \leq s$. In $W^{s,p}(\Omega)$ we define a norm by

$$\|u\|_{W^{s,p}(\Omega)}^p := \sum_{|\alpha| \leq s} \|D^\alpha u\|_{L^p(\Omega)}^p, \quad p < \infty,$$

$$\|u\|_{W^{s,\infty}(\Omega)} := \max_{|\alpha| \leq s} \|D^\alpha u\|_{L^\infty(\Omega)}.$$

If $p = 2$ we use the notation $W^{s,2}(\Omega) =: H^s(\Omega)$ and define an inner product by

$$(u,v)_{H^s(\Omega)} := \sum_{|\alpha| \leq s} \int_\Omega D^\alpha u(x) D^\alpha v(x) dx.$$

The closures of $C_0^\infty(\Omega)$ in $W^{s,p}(\Omega)$ and in $H^s(\Omega)$ are denoted by $W_0^{s,p}(\Omega)$ and $H_0^s(\Omega)$, respectively.

$W^{-s,p'}(\Omega)$ and $H^{-s}(\Omega)$ denote the dual spaces of $W_0^{s,p}(\Omega)$ and $H_0^s(\Omega)$, respectively. Here $s \in \mathbb{N}$ and $1 < p' < \infty$ such that $\frac{1}{p} + \frac{1}{p'} = 1$.

It is possible to define *fractional-order Sobolev space* $W^{s,p}(\Omega)$ for $s \in \mathbb{R}$ as the space of all $u \in W^{m,p}(\Omega)$ such that

$$\iint_{\Omega \times \Omega} \frac{|D^\alpha u(x_1) - D^\alpha u(x_2)|^p}{|x_1 - x_2|^{N+rp}} dx_x dx_2$$

is finite for $|\alpha| = m$, when $s = m + r$ and r is non-negative and not an integer. We define a norm by

$$\|u\|_{W^{s,p}(\Omega)} = \left(\|u\|_{W^{m,p}(\Omega)}^p + \sum_{|\alpha|=m} \iint_{\Omega \times \Omega} \frac{|D^\alpha u(x_1) - D^\alpha u(x_2)|^p}{|x_1 - x_2|^{N+rp}} dx_1 dx_2 \right)^{1/p}.$$

For more deeper study we refer to [23, 43].

8.5 Embedding and trace operator

Our main goal is the solution and control of PDE. We will see in Section 8.7 that solutions of PDEs are typically found as elements of Sobolev spaces. On the other hand, suitable boundary conditions are an essential part of PDE, which leads to determination of spaces of functions defined on Γ that contains a restriction (called *trace*) of elements of Sobolev spaces with respect to the boundary Γ.

8 Additional Material

For a general domain Ω the boundary can be extremely pathological. At the same time, requiring Γ to be smooth seems too restrictive, as surfaces with corners occur in numerous applications. The compromise addresses to Lipschitz regularity of the boundary Γ.

Definition 8.5.1. *A domain Ω is said to be <u>of class C^k</u>, $k \geqslant 1$, if every point on Γ has a neighborhood U so that $\Gamma \cap U$ is a C^k-surface and, moreover, $\Omega \cap U$ is "on one side" of Γ.*

Definition 8.5.2. *Let Ω be an open subset of $\mathbb{R}^N$. We say that its boundary Γ is of class $C^{k,1}$ if for every $x \in \Gamma$ there exists a neighborhood U of x in $\mathbb{R}^N$ and new orthogonal coordinates $\{y_1, \ldots, y_N\}$ such that*

(i) U is an hypercube in the new coordinates:

$$U = \{(y_1, \ldots, y_N) \mid -a_j < y_j < a_j, \ j = 1, \ldots, N\};$$

(ii) there exists a function φ of class $C^{k,1}$ defined in

$$U' = \{(y_1, \ldots, y_{N-1}) \mid -a_j < y_j < a_j, \ j = 1, \ldots, N-1\}$$

and such that

$$|\varphi(y')| \leqslant \frac{a_N}{2} \text{ for every } y' = (y_1, \ldots, y_{N-1}) \in U',$$
$$\Omega \cap U = \{y = (y', y_N) \in U \mid y_N < \varphi(y')\},$$
$$\Gamma \cap U = \{y = (y', y_N) \in U \mid y_N = \varphi(y')\}.$$

(φ belongs to the class $C^{k,1}$ if φ is k times continuously differentiable and its derivatives of order k are Lipschitz continuous)

In other words, in a neighborhood of x an open set Ω is below the graph of φ and consequently the boundary Γ is the graph of φ. In the following, we restrict ourselves to the case when Ω is bounded domain. We say that a bounded domain Ω has a <u>locally Lipschitz boundary</u>, if each point x on the boundary Γ has a neighborhood $U(x)$ whose intersection with Γ is a graph of a Lipschitz continuous function.

Note the Green's formula is valid in any bounded Lipschitz domain, as is shown in Nečas [32, Theorem 1.1].

Definition 8.5.3 (Embedding). *Let X and Y be normed spaces. X is <u>embedded</u> into Y, written $X \hookrightarrow Y$, if X is a subspace of Y and $\|u\|_Y \leqslant c \|u\|_X$ for all $u \in X$. The constant c is called the <u>embedding constant</u>.*

Sobolev's embedding theorem brings fundamental properties of the Sobolev spaces.

Theorem 8.5.4 (Sobolev's embedding theorem). *Let Ω be a bounded open subset of $\mathbb{R}^N$ with a Lipschitz boundary. For $s, t \in [0, \infty)$ and $p, q \in (1, \infty)$ the following inclusions hold*

(i) $W^{s,p}(\Omega) \hookrightarrow W^{t,q}(\Omega)$ *for* $s \geqslant t$, $q \geqslant p$ *such that* $s - \frac{N}{p} = t - \frac{N}{q}$

(ii) $W^{s,p}(\Omega) \hookrightarrow C^{k,\alpha}(\Omega)$ *for* $k < s - \frac{N}{p} < k+1$, $\alpha = s - k - \frac{N}{p}$, *and a non-negative integer* k.

If $0 < k \neq$ integer, the $W^{s,p}(\Omega) \hookrightarrow C^k(\Omega)$ is also valid for $k = s - \frac{N}{p}$.

Proof. For integer s, t and k see Grisvard [23, Theorem 1.4.4.1] or Adams [1, Theorem 4.12, Theorem 5.22]. For non-integer s, t and k we refer to Triebel [43, Theorem 4.6.1., Remark 4.6.1] or Bergh, Löfström [8, Theorem 6.5.1]. □

We outline below some embedding results for traces of $W^{s,p}(\Omega)$.

Theorem 8.5.5 (Trace theorem). *Let Ω be a bounded Lipschitz domain. There exists a linear continuous mapping $\tau : W^{1,p}(\Omega) \to L^p(\Gamma)$ such that for all $y \in C(\overline{\Omega})$*

$$(\tau y)(x) = y(x) \quad a.e. \text{ on } \Gamma$$

holds.

The element τy is called <u>trace</u> of y on Γ and the mapping τ is called <u>trace operator</u>.

Theorem 8.5.6 (A boundary trace embedding theorem). *Let Ω be a bounded Lipschitz domain. Suppose that $sN < p$ and $p \leqslant q \leqslant p^* = (N-1)p/(N-sp)$. Then $W^{s,p}(\Omega) \hookrightarrow L^q(\Gamma)$.*

Theorem 8.5.7. *Let Ω be a C^k-domain, $k \geqslant 1$ and $1 < p < \infty$. Then the trace operator of $W^{k,p}(\Omega)$ to $W^{k-1/p,p}(\Gamma)$ is continuous.*

Proof. See Adams [1, Theorem 7.53]. □

8.6 Nemyckii-operator

The analysis of nonlinear optimal control problems as well as of nonlinear PDEs leads to introduction of a class of nonlinear mappings. We also state most important results on the composition of $L^p(E)$ [1] with nonlinear functions. For a more detailed treatment, the reader could consult [29, 46].

---
[1] E is a bounded measurable set (for elliptic PDE: $E = \Omega$ or $E = \Gamma$)

8 Additional Material

Definition 8.6.1 (Nemyckii-operator). *Let $E \subset \mathbb{R}^N$ be a closed measurable set and $\phi = \phi(x,y) : E \times \mathbb{R} \to \mathbb{R}$ a real-valued function. A mapping Φ defined by*

$$\Phi(y) := \phi(\cdot, y(\cdot)),$$

where $y = y(x) : E \to \mathbb{R}$ and $z(x) = \phi(x, y(x)) : E \to \mathbb{R}$, is called <u>Nemyckii-operator</u> or <u>superposition-operator</u>.

Definition 8.6.2. *A function $\phi = \phi(x,y) : E \times \mathbb{R} \to \mathbb{R}$ is said to satisfy*

(i) *the <u>Carathéodory condition</u> if ϕ is measurable with respect to x for every fixed $y \in \mathbb{R}$ and continuous with respect to y for all $x \in E$.*

(ii) *the <u>boundedness condition</u> of order k if there exists a constant $K > 0$ such that*

$$|D_y^l \phi(x, 0)| \leqslant K$$

holds for all $x \in E$ and $l = 1, \ldots, k$.

(iii) *ϕ satisfies the <u>local Lipschitz condition</u> of order k, if for every constant $M > 0$ there exists a constant $L(M) > 0$ such that for almost all $x \in E$ the inequality*

$$|D_y^k \phi(x, y) - D_y^k \phi(x, z)| \leqslant L_\phi(M)|y - z|$$

holds for all $y, z \in [-M, M]$.

(iv) *ϕ is <u>local Lipschitz continuous</u> with respect to y if it satisfies the local Lipschitz condition according to (iii) with $k = 0$.*

Theorem 8.6.3. *Let $\phi = \phi(x,y)$ be a measurable function with respect to $x \in E$ and for every $y \in \mathbb{R}$, which is local Lipschitz continuous with respect to $y \in \mathbb{R}$ and satisfies the boundedness condition of order 0. Then the associated Nemyckii-operator is continuous in $L^\infty(E)$. Moreover, for $r \in [1, \infty]$ the inequality*

$$\|\Phi(x) - \Phi(y)\|_{L^r(E)} \leqslant \|x - y\|_{L^r(E)}$$

is valid for all functions $x, y \in L^\infty(E)$ with $\|x\|_{L^\infty(E)} \leqslant M$ and $\|y\|_{L^\infty(E)} \leqslant M$.

Lemma 8.6.4. *Let $\phi = \phi(x,y)$ be a measurable function with respect to $x \in E$ for all $y \in \mathbb{R}$ and for almost all $x \in E$ twice partial differentiable on y.*

If ϕ satisfies the boundedness and Lipschitz conditions of order $k = 2$, then the associated Nemyckii-operator Φ is twice differentiable in $L^\infty(E)$ and for arbitrary $M > 0$, the Lipschitz condition

$$\|\Phi_y(y_1) - \Phi_y(y_2)\|_{L^\infty(E)} \leqslant L_\Phi(M) \|y_1 - y_2\|_{L^\infty(E)}$$

holds for all $y_i \in L^\infty(E)$ such that $\|y_i\|_{L^\infty(E)} \leqslant M$, $i = 1, 2$. In particular,

$$\|\Phi_y(y)\|_{L^\infty(E)} \leqslant K_\Phi$$

holds for all $y \in L^\infty(E)$ such that $\|y\|_{L^\infty(E)} \leqslant M$. The same properties, with different constants, are valid for $\Phi_{yy}(\cdot)$.

Proof. See [46, Lemma 4.10, Satz 4.20]. □

Note that $\Phi_y(y)$ and $\Phi_{yy}(y)$ defined by

$$\Phi_y(y)[h](\cdot) := \phi_y(\cdot, y(\cdot)) \, h(\cdot),$$
$$\Phi_{yy}(y)[h_1, h_2](\cdot) := \phi(\cdot, y(\cdot)) \, h_1(\cdot) h_2(\cdot)$$

present first and second Frèchet derivative of $\Phi(y)$, respectively.

8.7 Elliptic PDEs with Neumann boundary condition

In this section we collect some preparatory materials concerning elliptic partial differential equations (PDEs) needed in the thesis.

Let Ω be a bounded domain with Lipschitz boundary Γ.
First we define the elliptic operator $\mathcal{A} : H^1(\Omega) \to H^1(\Omega)^*$ as

$$\mathcal{A} y(v) = ((\nabla v)^\top, A_0 \nabla y)_\Omega + (cy, v)_\Omega.$$

A_0 is an $N \times N$ symmetric matrix with Lipschitz continuous entries on $\overline{\Omega}$ such that $\rho^\top A_0(\xi) \rho \geqslant m_0 |\rho|^2$ holds with some $m_0 > 0$ for all $\rho \in \mathbb{R}^N$ and almost all $\xi \in \overline{\Omega}$.

The adjoint operator will be denoted by $\mathcal{A}^*$, i.e.,

$$\mathcal{A}^* v(y) = ((\nabla y)^\top, A_0^\top \nabla v)_\Omega + (cv, y)_\Omega.$$

8 Additional Material

The corresponding <u>*co-normal derivatives*</u> are

$$\partial_n y := \frac{\partial y}{\partial \nu_{\mathcal{A}}} = \vec{n}^\top A_0 \nabla y, \qquad \partial_n^* v := \frac{\partial v}{\partial \nu_{\mathcal{A}}^*} = \vec{n}^\top A_0^\top \nabla v.$$

Lemma 8.7.1. *For every* $y \in W^{2,p}(\Omega)$ *and* $v \in W^{2,p'}(\Omega)$ *with* $\frac{1}{p} + \frac{1}{p'} = 1$, *there holds*

$$\int_\Omega \mathcal{A} y \, v \, dx - \int_\Omega y \, \mathcal{A}^* \, v \, dx = \int_\Gamma \partial_n y \, v \, ds - \int_\Gamma y \, \partial_n^* v \, ds.$$

The elliptic partial differential equation with Neumann boundary condition one presents in general form by

$$\begin{aligned} \mathcal{A} y + d(y) &= r_1 \quad \text{in } \Omega, \\ \partial_n y + b(y) &= r_2 \quad \text{on } \Gamma \end{aligned} \tag{8.7.1}$$

The function $y \in C^2(\Omega) \cap C^1(\overline{\Omega})$ is called a *classical solution* of (8.7.1) if it together with its derivatives satisfies equation in Ω and the boundary condition pointwise on Γ. If $y \in W^{2,p}(\Omega)$, its derivatives in the equation are understood in the sense of distributions, and if the boundary condition are interpreted in the sense of traces, then y is called a *strong solution*.

However, in many practically important applications the regularity properties of the coefficient functions or of the data are too weak to ensure the existence of such solutions. That is why one reasons to implement and use so-called *weak solution*.

Definition 8.7.2. *A function* y *is called a <u>weak solution</u> of the state equation* (8.7.1), *if* $y \in H^1(\Omega) \cap C(\overline{\Omega})$ *and*

$$a[y, v] + (d(y), v)_\Omega + (b(y), v)_\Gamma = (r_1, v)_\Omega + (r_2, v)_\Gamma \tag{8.7.2}$$

holds for all $v \in H^1(\Omega)$.

A strikingly useful tool for the existence of (unique) weak solutions of elliptic PDEs is provided by Lax-Milgram Theorem, see e.g. [28, Theorem 1.19.3]. The basic ingredients in the proof are coercivity and continuity assumptions on the bilinear form $a[\cdot, \cdot]$ and the Riesz Representation Theorem, see e.g. [28, Theorem 1.19.2]. Results concerning the existence and regularity of weak solution with respect to various data are discussed in the following section.

Another useful property of the elliptic equation is the *maximum principle*, see Lemma 8.7.3 below, which asserts that the solution of the elliptic equation cannot have a maximum (or a minimum) in the interior of the domain where they are defined, see e.g. [12, 35]. The maximum principle can be used to show that solutions of certain equations must be non-negative.

Lemma 8.7.3 (Maximum principle). *Assume that $\mathcal{A}y \geqslant 0$ (or, respectively, $\mathcal{A}y \leqslant 0$) in a bounded domain Ω and $c = 0$ in Ω. Then the maximum (or, respectively, the minimum) of y is achieved on Γ.*

Proof. See [33, Theorem A2.8.]. □

Corollary 8.7.4. *Let Ω be a bounded and $c \leqslant 0$. If $\mathcal{A}y = \mathcal{A}v$ in Ω and $y = v$ on Γ, then $y = v$ in Ω. If $\mathcal{A}y \leqslant \mathcal{A}v$ in Ω and $y \geqslant v$ on Γ, then $y \geqslant v$ in Ω.*

8.8 Regularity of the solution

Let Ω be a bounded domain in $\mathbb{R}^N$ with a $C^{1,1}$ boundary Γ.

Linear case.

Let us consider the linear elliptic equation

$$\begin{aligned} \mathcal{A}y + a_1 y &= r_1 \quad \text{in } \Omega, \\ \partial_n y + b_1 y &= r_2 \quad \text{on } \Gamma \end{aligned} \tag{8.8.1}$$

with non-negative constans a_1 and b_1 that do not equal to zero simultaneously.

Lemma 8.8.1. *[22, Lemma 2.4] Let Assumption **(A1)** be satisfied. For every $r_1 \in L^N(\Omega)$ and $r_2 \in L^p(\Gamma)$ with $p \in (\frac{2N-2}{N}, \infty)$, the linear PDE (8.8.1) has a unique weak solution $y \in W^{1,p}(\Omega)$. Moreover, the a priori estimate*

$$\|y\|_{W^{1,p}(\Omega)} \leqslant C_{p,\Omega} \left(\|r_1\|_{W^{1,p'}(\Omega)^*} + \|r_2\|_{W^{-1/p,p}(\Gamma)} \right)$$

holds with a constant $C_{p,\Omega} > 0$ independent of r_1, r_2.

Theorem 8.8.2. *[23, Theorem 2.2.2.3] Under the assumption of the previous theorem, the linear PDE (8.8.1) possesses a unique weak solution $y \in H^2(\Omega)$ for any given $r_1 \in L^2(\Omega)$ and $r_2 \in L^2(\Gamma)$. It satisfies the a priori estimate*

$$\|y\|_{H^2(\Omega)} \leqslant C_\Omega \|r_1\|_{L^2(\Omega)} + \|r_2\|_{L^2(\Gamma)}$$

with a positive constant C_Ω independent of r_1, r_2.

8 Additional Material

Nonlinear case.

By studying of the nonlinear PDE (8.7.1) one has to assume the nonlinear functions $d(\cdot)$ and $b(\cdot)$ to be twice differentiable, monotone, locally bounded and locally Lipschitz continuous.

Theorem 8.8.3. *[14, Theorem 4.3]*
Let Assumptions **(A1)**–**(A2)** *be satisfied. For any given $r_1 \in \mathcal{M}(\Omega)$ and $r_2 \in \mathcal{M}(\Gamma)$, there exists a unique weak solution of (8.7.1) in the space $W^{1,p}(\Omega)$ for every $p \in [1, \frac{N}{N-1})$, which satisfies the a priori estimate*

$$\|y\|_{W^{1,p}(\Omega)} \leqslant C_{p,\Omega}(\|r_1\|_{\mathcal{M}(\Omega)} + \|r_2\|_{\mathcal{M}(\Gamma)}). \tag{8.8.2}$$

Theorem 8.8.4. *[46, Theorem 7.1]*
Let Assumptions **(A1)**–**(A2)** *be satisfied. For any given right hand side $r_1 \in L^2(\Omega)$ and $r_2 \in L^2(\Gamma)$ there exists a unique weak solution of (8.7.1) in the space $H^1(\Omega)$. It satisfies the a priori estimate*

$$\|y\|_{H^1(\Omega)} \leqslant C_\Omega(\|r_1\|_{L^2(\Omega)} + \|r_2\|_{L^2(\Gamma)}). \tag{8.8.3}$$

Lemma 8.8.5. *[46, Theorem 7.3]*
Let Assumptions **(A1)**–**(A2)** *be satisfied. For any given right hand side $r_1 \in L^r(\Omega)$ and $r_2 \in L^q(\Gamma)$ with $r > \frac{N}{2}$ and $q > N-1$, there exists a unique weak solution of (8.7.1) in the space $H^1(\Omega) \cap C(\bar{\Omega})$. It satisfies the a priori estimate*

$$\|y\|_{H^1(\Omega)} + \|y\|_{C(\bar{\Omega})} \leqslant C_\Omega(\|r_1\|_{L^r(\Omega)} + \|r_2\|_{L^q(\Gamma)}). \tag{8.8.4}$$

Lemma 8.8.6. *Under assumptions of the previous lemma, for every $r_1 \in L^N(\Omega)$ and $r_2 \in L^\infty(\Omega)$ the unique weak solution of (8.7.1) belongs to $W^{1,p}(\Omega)$ for all $p \in [1, \infty)$.*

Proof. We rewrite (8.7.1) in the form

$$\mathcal{A}y + d_y(y) = r_1 + d_y(y)y - d(y) \quad \text{in } \Omega,$$
$$\partial_n y + b_y(y) = r_2 + b_y(y) - b(y) \quad \text{on } \Gamma.$$

The right hand sides are elements of $L^N(\Omega)$ and $L^\infty(\Gamma)$, respectively, and the claim follows from Lemma 8.8.1. $\square$

8.9 Dirichlet boundary problem

Let Ω be a bounded domain in $\mathbb{R}^N$ with a $C^{1,1}$ boundary Γ. We consider the homogeneous Dirichlet problem [1]

$$\begin{aligned}\mathcal{A}y + d(y) &= r_1 \quad \text{in } \Omega, \\ y &= 0 \quad \text{on } \Gamma.\end{aligned} \qquad (8.9.1)$$

The elliptic operator $\mathcal{A}: H_0^1(\Omega) \to H_0^1(\Omega)^*$ is defined as $\mathcal{A}y(v) = a[y,v]$, where

$$a[y,v] = ((\nabla v), A_0 \nabla y)_\Omega + (a_1^\top \nabla y, v)_\Omega + (a_0 y, v)_\Omega.$$

and satisfied Assumption (**A1'**). Analogous to Definition 8.7.2 we say that y is a *weak solution* of (8.9.1), if $y \in H_0^1(\Omega) \cap C(\overline{\Omega})$ and

$$a[y,v] + (d(y), v)_\Omega = (r_1, v)_\Omega$$

holds for all $v \in H_0^1(\Omega)$.

Theorem 8.9.1. *[13, Theorem 4]*
Let Assumptions (**A1'**)–(**A2'**) *be satisfied. For any given measurable function r_1 in Ω there exists a unique weak solution of (8.9.1) in the space $W_0^{1,p}(\Omega)$ for every $p \in [1, \frac{N}{N-1})$, which satisfies the a priori estimate*

$$\|y\|_{W_0^{1,p}(\Omega)} \leqslant C_{p,\Omega} \|r_1\|_{M(\Omega)}.$$

Lemma 8.9.2. *Under assumptions of the previous lemma, for every $r_1 \in L^N(\Omega)$ the unique weak solution of (8.7.1) belongs to $W_0^{1,p}(\Omega)$ for all $p \in [1, \infty)$.*

Proof. We rewrite (8.9.1) in the form

$$\begin{aligned}\mathcal{A}y + d_y(y) &= r_1 + d_y(y)y - d(y) \quad \text{in } \Omega, \\ y &= 0 \quad \text{on } \Gamma.\end{aligned}$$

The right hand side is an element of $L^N(\Omega)$ and the claim follows from Lemma 8.9.4 below. $\square$

Lemma 8.9.3. *Under Assumptions* (**A1'**)–(**A2'**) *and for any given $r_1 \in L^2(\Omega)$, the semilinear equation (8.9.1) possesses a unique weak solution $y \in H_0^1(\Omega) \cap H^2(\Omega)$. It satisfies the a priori estimate*

$$\|y\|_{H^1(\Omega)} + \|y\|_{L^\infty(\Omega)} \leqslant C_\Omega \left(\|r_1\|_{L^2(\Omega)} + 1\right)$$

---
[1] In present work we study only homogeneous Dirichlet boundary condition.

8 Additional Material

with a constant C_Ω independent of r_1.

Proof. The existence and uniqueness of a weak solution $y \in H_0^1(\Omega) \cap L^\infty(\Omega)$ is a standard result [46, Theorem 4.8]. It satisfies

$$\|y\|_{H^1(\Omega)} + \|y\|_{L^\infty(\Omega)} \leqslant C_\Omega \left(\|r_1\|_{L^2(\Omega)} + 1\right) =: M$$

with some constant C_Ω independent of u. Lemma 8.6.4 implies that $d(y) \in L^\infty(\Omega)$. Using the embedding $L^\infty(\Omega) \hookrightarrow L^2(\Omega)$, we conclude that the difference $r_1 - d(y)$ belongs to $L^2(\Omega)$. Owing to assumption (A1), $y \in H^2(\Omega)$ follows from Theorem 8.8.2. □

In the text we also need the corresponding regularity result for the linear equation, i.e.,

$$\begin{aligned} \mathcal{A}y + a_1 y &= r_1 \quad \text{in } \Omega, \\ y &= 0 \quad \text{on } \Gamma, \end{aligned} \tag{8.9.2}$$

where $a_1 \in L^\infty(\Omega)$ is non-negative.

Theorem 8.9.4. *(see [30]) Let Assumption* (**A1′**) *be satisfied. For every $r_1 \in L^N(\Omega)$ the linear PDE (8.9.2) has a unique weak solution $y \in W_0^{1,p}(\Omega)$. Morover, the a priori estimate*

$$\|y\|_{W_0^{1,p}(\Omega)} \leqslant C_{p,\Omega} \|r_1\|_{W^{1,p'}(\Omega)^*}$$

holds with a constant $C_{p,\Omega} > 0$ independent of r_1, r_2.

Theorem 8.9.5. *[23, Theorem 2.2.2.3] Under Assumption* (**A1′**) *the linear equation (8.9.2) possesses a unique weak solution $y \in H_0^1(\Omega) \cap H^2(\Omega)$ for any given $r_1 \in L^2(\Omega)$. It satisfies the a priori estimate*

$$\|y\|_{H^2(\Omega)} \leqslant C_\Omega \|r_1\|_{L^2(\Omega)}$$

with a constant C_Ω independent of u.

9 Assumptions

Throughout present work, if nothing else is mentioned, Ω is a bounded domain in $\mathbb{R}^N$, $N = \{2,3\}$, which has a $C^{1,1}$ boundary Γ.

Assumption for (P)

(A1) The operator $\mathcal{A}: H^1(\Omega) \to H^1(\Omega)^*$ is defined as $\mathcal{A}y(v) = a[y,v]$, where

$$a[y,v] = ((\nabla v)^\top, A_0 \nabla y)_\Omega + (a_0 y, v)_\Omega.$$

A_0 is an $N \times N$ symmetric matrix with Lipschitz continuous entries on $\overline{\Omega}$ such that $\rho^\top A_0(\xi)\rho \geqslant m_0 |\rho|^2$ holds with some $m_0 > 0$ for all $\rho \in \mathbb{R}^N$ and almost all $\xi \in \overline{\Omega}$. Moreover, $a_0 \in L^\infty(\Omega)$. The symbol ∂_n denotes the co-normal derivative associated to A_0.

The bilinear form $a[\cdot,\cdot]$ is assumed to be continuous and coercive, i.e.,

$$a[y,v] \leqslant \overline{c}\,\|y\|_{H^1(\Omega)}\,\|v\|_{H^1(\Omega)},$$
$$a[y,y] \geqslant \underline{c}\,\|y\|^2_{H^1(\Omega)}$$

for all $y, v \in H^1(\Omega)$ with some positive constants $\overline{c}$ and $\underline{c}$. (This is satisfied if ess inf $a_0 > 0$.) The right hand side f is taken from $L^N(\Omega)$.

(A2) The functions $d(\xi,y)$ and $b(\xi,y)$ belong to the class of C^2 with respect to y for almost all $\xi \in \Omega$ or $\xi \in \Gamma$, respectively. Moreover, d_{yy} and b_{yy} are assumed to be a locally bounded and locally Lipschitz-continuous functions with respect to y, i.e., the following conditions hold true: there exist $K_d > 0$ and $K_b > 0$ such that

$$|d(\xi,0)| + |d_y(\xi,0)| + |d_{yy}(\xi,0)| \leqslant K_d,$$
$$|b(\xi,0)| + |b_y(\xi,0)| + |b_{yy}(\xi,0)| \leqslant K_b,$$

and for any $M > 0$, there exist $L_d(M) > 0$ and $L_b(M) > 0$ such that

$$|d_{yy}(\xi,y_1) - d_{yy}(\xi,y_2)| \leqslant L_d(M)\,|y_1 - y_2| \quad \text{a.e. in } \Omega,$$
$$|b_{yy}(\xi,y_1) - b_{yy}(\xi,y_2)| \leqslant L_b(M)\,|y_1 - y_2| \quad \text{a.e. on } \Gamma$$

9 Assumptions

for all $y_1, y_2 \in \mathbb{R}$ satisfying $|y_1|, |y_2| \leq M$.

Additionally for all $y \in \mathbb{R}$ we assume $d_y(\xi, y) \geq 0$ a.e. in Ω and $b_y(\xi, y) \geq 0$ for almost all $\xi \in \Gamma$.

(A3) The function $\psi(\xi, y, u)$ is measurable with respect to $\xi \in \Gamma$ for each y and u, and of class C^2 with respect to y and u for almost all $\xi \in \Gamma$. Again the second derivatives are assumed to be locally bounded and locally Lipschitz-continuous functions, i.e., the following conditions hold: there exists $K_\psi > 0$ such that

$$|\psi(\xi, 0, 0)| + |\psi_u(\xi, 0, 0)| + |\psi_y(\xi, 0, 0)|$$
$$+ |\psi_{uu}(\xi, 0, 0)| + |\psi_{yu}(\xi, 0, 0)| + |\psi_{yy}(\xi, 0, 0)| \leq K_\psi$$

and for any $M > 0$, there exists $L_\psi(M) > 0$ such that

$$|\psi_{yy}(\xi, y_1, u_1) - \psi_{yy}(\xi, y_2, u_2)| \leq L_\psi(M)(|y_1 - y_2| + |u_1 - u_2|),$$
$$|\psi_{yu}(\xi, y_1, u_1) - \psi_{yu}(\xi, y_2, u_2)| \leq L_\psi(M)(|y_1 - y_2| + |u_1 - u_2|),$$
$$|\psi_{uy}(\xi, y_1, u_1) - \psi_{uy}(\xi, y_2, u_2)| \leq L_\psi(M)(|y_1 - y_2| + |u_1 - u_2|),$$
$$|\psi_{uu}(\xi, y_1, u_1) - \psi_{uu}(\xi, y_2, u_2)| \leq L_\psi(M)(|y_1 - y_2| + |u_1 - u_2|)$$

for all $y_i, u_i \in \mathbb{R}$ satisfying $|y_i|, |u_i| \leq M$, $i = 1, 2$. Analogous conditions are assumed to hold for $g_i = g_i(\xi, y, u)$, $i = 1, ..., s$ and $\phi = \phi(\xi, y)$.

(A4) There exist $\tau > 0$ and $\hat{u} \in L^\infty(\Gamma)$ such that

$$g_i(y^*, u^*) + g_{i,y}(y^*, u^*)\hat{y} + g_{i,u}(y^*, u^*)\hat{u} \leq -\tau \text{ a.e. on } \Gamma$$

holds a.e. on Γ, where $\hat{y} \in W^{1,\bar{p}}(\Omega)$ is the unique solution of the linearized PDE

$$\mathcal{A}\hat{y} + d_y(y^*)\hat{y} = f \quad \text{in } \Omega,$$
$$\partial \hat{y} + b_y(y^*)\hat{y} = \hat{u} \quad \text{on } \Gamma.$$

(A5) There is a constant $m > 0$ such that the properties

$$\psi_{uu}(\xi, y, u) \geq m \quad \forall \xi \in \Gamma,\ (y, u) \in \mathbb{R}^2,$$
$$\text{and} \quad g_{i,u}(\xi, y, u) \geq m \quad \forall \xi \in \Gamma,\ (y, u) \in \mathbb{R}^2$$

hold.

(**A6**) Suppose that $S_i \cap S_j = \emptyset$ for all $i,j = 1,\ldots,s$, $i \neq j$, where
$$S_i := \{\xi \in \Gamma \ : \ -\sigma_i \leqslant g_i(y^*, u^*) \leqslant 0\}.$$

Moreover, we assume that the boundary value problem
$$\mathcal{A}^* p + d_y(y^*) p = r_1 \quad \text{in } \Omega,$$
$$\partial^* p + [b_y(y^*) + \sum_{i=1}^{s} \chi_{S_i} g_{i,u}^{-1}(y^*, u^*) g_{i,y}(y^*, u^*)] p = r_2 \quad \text{on } \Gamma$$

has a unique weak solution $p \in H^1(\Omega)$ for all right hand sides $r_1 \in L^2(\Omega)$ and $r_2 \in L^2(\Gamma)$.

(**A7**) There exists a constant $\alpha > 0$ such that
$$\mathcal{L}_{xx}(w^*)(\delta x, \delta x) \geqslant \alpha \|\delta x\|_{L^2(\Omega) \times L^2(\Gamma)}^2$$

for all $\delta x = (\delta y, \delta u) \in W^{1,\bar{p}}(\Omega) \times L^\infty(\Gamma)$ satisfying the linearized PDE
$$\mathcal{A} \, \delta y + d_y(y^*) \, \delta y = 0 \quad \text{in } \Omega,$$
$$\partial_n \delta y + b_y(y^*) \, \delta y = \delu u \quad \text{on } \Gamma.$$

Assumption for (P′)

(**A1′**) The operator $A : H_0^1(\Omega) \to H_0^1(\Omega)^*$ is defined as $A\, y(v) = a[y, v]$, where
$$a[y, v] = ((\nabla v), A_0 \nabla y)_\Omega + (a_1^\top \nabla y, v)_\Omega + (a_0 y, v)_\Omega.$$

A_0 is an $N \times N$ matrix with Lipschitz continuous entries on $\overline{\Omega}$ such that $\rho^\top A_0(\xi)\, \rho \geqslant m_0\, |\rho|^2$ holds with some $m_0 > 0$ for all $\rho \in \mathbb{R}^N$ and almost all $\xi \in \overline{\Omega}$. Moreover, $a_1 \in L^\infty(\Omega)^N$ and $a_0 \in L^\infty(\Omega)$. The bilinear form $a[\cdot, \cdot]$ is not necessarily symmetric but it is assumed to be continuous and coercive, i.e.,
$$a[y, v] \leqslant \bar{c} \|y\|_{H^1(\Omega)} \|v\|_{H^1(\Omega)},$$
$$a[y, y] \geqslant \underline{c} \|y\|_{H^1(\Omega)}^2$$

for all $y, v \in H_0^1(\Omega)$ with some positive constants $\bar{c}$ and $\underline{c}$. A simple example is $a[y, v] = (\nabla y, \nabla v)$, corresponding to $A = -\Delta$.

(**A2′**) The function $d(\xi, y)$ is assumed to satisfy (**A2**).

(**A3′**) The function $\psi = \psi(\xi, y, u)$ satisfies conditions analogous to (**A3**) on Ω.

(**A4'**) There exist $\tau > 0$ and $\hat{u} \in L^\infty(\Omega)$ such that
$$g_i(y^*, u^*) + g_{i,y}(y^*, u^*)\hat{y} + g_{i,u}(y^*, u^*)\hat{u} \leqslant -\tau \quad \text{a.e. in } \Omega$$
holds a.e. on Γ, where $\hat{y} \in Y$ is the unique solution of the linearized PDE
$$\mathcal{A}\hat{y} + d_y(y^*)\hat{y} = \hat{u} \quad \text{in } \Omega,$$
$$\hat{y} = 0 \quad \text{on } \Gamma.$$

(**A5'**) Functions ψ and g_i satisfy Assumption (**A5**) within Ω.

(**A6'**) Suppose that $S_i' \cap S_j' = \emptyset$ for all $i, j = 1, \ldots, s$, $i \neq j$, where
$$S_i' := \{\xi \in \Omega \ : \ -\sigma_i \leqslant g_i(y^*, u^*) \leqslant 0\}.$$
Moreover, we assume that the boundary value problem
$$\mathcal{A}^*p + \Big[d_y(y^*) + \sum_{i=1}^s \chi_{S_i} g_{i,u}^{-1}(y^*, u^*) g_{i,y}(y^*, u^*)\Big]p = r_1 \quad \text{in } \Omega,$$
$$p = 0 \quad \text{on } \Gamma$$
has a unique weak solution $p \in H_0^1(\Omega)$ for all right hand sides $r_1 \in L^2(\Omega)$.

(**A7'**) There exists a constant $\alpha > 0$ such that
$$\mathcal{L}_{xx}(w^*)(\delta x, \delta x) \geqslant \alpha \|\delta x\|^2_{[L^2(\Omega)]^2}$$
for all $\delta x = (\delta y, \delta u) \in Y \times L^\infty(\Omega)$ satisfying the linearized PDE
$$\mathcal{A}\,\delta y + d_y(y^*)\,\delta y = \delta u \quad \text{in } \Omega,$$
$$\delta y = 0 \quad \text{on } \Gamma.$$

Notations

General notations

Ω	a bounded domain in $\mathbb{R}^N$, $N = \{2,3\}$
Γ	the boundary of Ω
X, Y	Banach spaces with the norm $\|\cdot\|_X$
$X \times Y$	Cartesian product of X and Y
X^s	$:= \underbrace{X \times \cdots \times X}_{s\,times}$
2^X	the set of subsets of X
$B_r^X(x^*)$	the open ball of radius $r > 0$ centered at $x^* \in X$, $B_r^X(x^*) := \{x \in X : \|x - x^*\|_X < r\}$
$f : X \to \mathbb{R}$	a functional defined on X
X^*	the dual space to X, i.e., the space of linear continuous functionals f with $\|f\|_{X^*} := \sup_{\|u\|_X=1} |f(u)|$
$(\cdot, \cdot)_\Omega$	the inner product in $L^2(\Omega)$
$(\cdot, \cdot)_\Gamma$	the inner product in $L^2(\Gamma)$
$\langle \cdot, \cdot \rangle_{X,Y}$	a dual product with first component in X and the second component in Y
$a[\cdot, \cdot]$	a bilinear form
$A : X \to Y$	a mapping from X to Y
$A^* : Y^* \to X^*$	the adjoint operator of the continuous linear operator A, i.e., $\langle A^*\lambda, x \rangle_{X^*,X} = \langle \lambda, Ax \rangle_{Y^*,Y}$, for all $x \in X$ and $\lambda \in Y^*$

Definition of classical spaces

$L^p(\Omega)$	the set of all measurable functions having finite norm $\|f\|_{L^p(\Omega)} = \left(\int_\Omega |f(x)|^p dx\right)^{\frac{1}{p}}$
$L^\infty(\Omega)$	the set of all measurable functions with a finite norm $\|f\|_{L^\infty(\Omega)} = \operatorname{ess\,sup}_{x \in \Omega} |f(x)| := \inf_{\mu(F)=0} \sup_{x \in E \setminus F} |f(x)|$
$L^1_{loc}(\Omega)$	the space of functions integrable on each compact subset of Ω

$C^0(\Omega)$ — the Banach space of continuous functions $f : \Omega \to \mathbb{R}$ with
$$\|f\| = \sup_{x \in \Omega} |f(x)|$$

$C^k(\Omega)$ — the space of all functions, which have continuous partial derivatives up to order $k \in \mathbb{N}$

$C^\infty(\Omega)$ — the space of all functions belonging to $C^k(\Omega)$, for any k

$C_0^k(\Omega)$ — the space of functions k-times continuous differentiable on Ω and have a compact support, $\mathrm{supp} f = \overline{\{x \in \Omega : f(x) = 0\}}$

$C_0^\infty(\Omega)$ — all functions from $C^\infty(\Omega)$ with a compact support

$C^{k,\alpha}(\Omega)$ — the space of function $f \in C^k(\Omega)$ which has a locally Hölder continuous derivative of k-th order with an exponent α
$$\|f\|_{C^{k,\alpha}(\Omega)} = \sup_{x \ne y \in \Omega} \frac{|f(x)-f(y)|}{|x-y|^\alpha}, \ 0 < \alpha \leqslant 1$$

$W^{s,p}(\Omega)$ — the Sobolev space
$$W^{s,p}(\Omega) := \begin{cases} \{f \in L^p(\Omega) \mid D^\alpha f \in L^p(\Omega), |\alpha| \leqslant s, \ s \text{ is an integer}\}, \\ \{f \in W^{m,p}(\Omega) \mid \iint_{\Omega \times \Omega} \frac{|D^\alpha f(x) - D^\alpha f(y)|^p}{|x-y|^{N+rp}} dx\, dy < \infty, \\ |\alpha| = m, \text{ when } s = m+r \text{ and } 0 < r \ne \text{integer}\} \end{cases}$$

with the norm
$$\|f\|_{W^{s,p}(\Omega)} = \begin{cases} \left(\sum_{|\alpha| \leqslant s} \int_\Omega |D^\alpha f|^p dx\right)^{\frac{1}{p}}, \ s \text{ is an integer} \\ \left(\|f\|_{W^{m,p}(\Omega)}^p + \sum_{|\alpha|=m} \iint_{\Omega \times \Omega} \frac{|D^\alpha f(x) - D^\alpha f(y)|^p}{|x-y|^{N+rp}} dx\, dy\right)^{1/p}, \\ s = m+r \text{ and } 0 < r \ne \text{integer} \end{cases}$$

$W_0^{s,p}(\Omega)$ — the Sobolev space of all functions $f \in W^{s,p}(\Omega)$ with $f\big|_\Gamma = 0$

$W^{-s,p'}(\Omega)$ — the dual space to $W_0^{s,p}(\Omega)$, where $\frac{1}{p} + \frac{1}{p'} = 1$

$H^s(\Omega)$ — the Hilbert space $H^1(\Omega) = W^{s,2}(\Omega)$

$H_0^s(\Omega)$ — the space of all functions $f \in H^s(\Omega)$ with $f\big|_\Gamma = 0$

χ_D — the indicator function of the set D, i.e., $\chi_D = \begin{cases} 1, & x \in D \\ 0, & x \notin D \end{cases}$

Derivatives

$f_{x_i}(x) := \frac{\partial f}{\partial x_i}$ — the partial derivative of f with respect to x_i at $x = (x_1, \ldots, x_n) \in \mathbb{R}^n$

$\nabla f(x) = \left(\frac{\partial f}{\partial x_1}, \ldots, \frac{\partial f}{\partial x_n}\right)$ — the gradient of the function $f : \mathbb{R}^n \to \mathbb{R}$ at the point $x \in \mathbb{R}^n$

$\nabla^2 f(x) = \left[\frac{\partial^2 f}{\partial x_i \partial x_j}\right]_{i,j=1}^n$ — the Hessian matrix of second order partial derivatives of $f : \mathbb{R}^n \to \mathbb{R}$ at the point $x \in \mathbb{R}^n$

$D^\alpha y := \frac{\partial^{|\alpha|} y}{\partial x_1^{\alpha_1} \ldots \partial x_n^{\alpha_n}}$ — the partial derivative of order α
$|\alpha| = \sum_i \alpha_i, \ \alpha := (\alpha_1, \ldots, \alpha_n), \ \alpha_i \geqslant 0$

$\Delta y = \sum_{i=1}^{n} \frac{\partial^2 y}{\partial x_i^2}$ the Laplace operator of the function y

$\mathcal{A}$ an elliptic operator

$\partial_n y = \frac{\partial y}{\partial \nu_\mathcal{A}}$ the co-normal derivative of y with respect to $\mathcal{A}$, p.96

Space setting

Y equal either $W^{1,\bar{p}}(\Omega)$ or $H_0^1(\Omega) \cap H^2(\Omega)$

$X := W^{1,\bar{p}}(\Omega) \times L^\infty(\Gamma)$

$X' := Y \times L^\infty(\Omega)$

$W := W^{1,\bar{p}}(\Omega) \times L^\infty(\Gamma) \times W^{1,\bar{p}}(\Omega) \times [L^\infty(\Gamma)]^s$

$W_0 := H^1(\Omega) \times L^2(\Gamma) \times H^1(\Omega) \times [L^2(\Gamma)]^s$

$W^\infty := L^\infty(\Omega) \times L^\infty(\Gamma) \times L^\infty(\Omega) \times [L^\infty(\Gamma)]^s$

$W' := Y \times L^\infty(\Omega) \times Y \times [L^\infty(\Omega)]^s$

$Z := W^{1,\bar{p}'}(\Omega)^* \times L^\infty(\Gamma) \times W^{1,\bar{p}'}(\Omega)^* \times [L^\infty(\Gamma)]^s$

$Z_0 := H^1(\Omega)^* \times L^2(\Gamma) \times H^1(\Omega)^* \times [L^2(\Gamma)]^s$

$Z' := H_0^1(\Omega) \times L^2(\Omega) \times H_0^1(\Omega) \times [L^2(\Omega)]^s$

Notations for optimal control problems

y a state

u a control

p an adjoint state

μ_i a Lagrange multiplier associated to $i-$th inequality

$x := (y, u)$

$w := (y, u, p, \mu_1, \ldots, \mu_s)$

$\delta := (\delta_1, \ldots, \delta_s)$ a perturbation parameter

$J(x)$ an objective function

$J_x(x) = (J_y(x), J_u(x))^\top$

$\mathcal{L}(w)$ a Lagrange functional $\mathcal{L}: W \to \mathbb{R}$

$\mathcal{L}_x(w^*)$ partial derivative of the Lagrange function $\mathcal{L}$ with respect to variable $x \in X$ at the point $w^* \in W$

S_i a security set of level σ_i

$\mathcal{A}_i$ an active set corresponding to the i-th inequality

$L, L_1, \ldots$ Lipschitz constants

c_Ω an a priori constant

c_∞ an embedding constant w.r.t. $L^\infty-$spaces

$c_1, c_2, \ldots$ some positive constants

Index

adjoint state, 21
algorithm
    PDAS, 67
    SQP, 71

bilinear form, 87
bilinear form
    coercive, 88
    elliptic, 88
    positive semidefinite, 88
    symmetric, 88
boundary condition
    Dirichlet, 99
    Neumann, 96

co-normal derivative, 96
compact support, 90
condition
    boundedness, 94
    Carathéodory, 94
    Karush-Kuhn-Tucker, 11
    Legendre-Clebsch, 27
    Lipschitz, 58
    local Lipschitz, 94
    of optimality
        first-order, 11, 20
        second-order, 26
    regularity, 10
    second-order growth, 19
    Slater, 10, 20
constraint, 7
control, 6
control

boundary, 6
    distributed, 6
convex
    function, 87
    set, 86
    strictly, 87
cost functional, 6

distribution, 90
dual cone, 54
dual variable, 10

embedding, 92
equation
    adjoint state, 21
    generalized, 54
    gradient, 21
    linearized generalized, 57

inequality
    Hölder's, 86
    maximum, 85
    Young's, 85
inner product, 91

Lagrange function, 10
Lagrange functional, 20
Lagrange multiplier, 10, 21
Lagrangian, 10
Lipschitz
    boundary, 92
local Lipschitz continuous, 94
local quadratic convergence, 64

111

Maximum principle, 97
measure
    countable additive, 89
measure
    purely finite additive, 89
method
    FEM, 71
    generalized Newton, 53
    primal-dual active set, 67
    SQP, 13, 53
multi-index, 90

objective, 6
operator
    elliptic, 95
    Nemyckii-, 93
    trace, 93
optimal control problem, 6
optimal control problem
    auxiliary linear-quadratic, 37
    boundary control, 7, 8
    distributed control, 8
    linear-quadratic, 35

partial derivative, 90
point
    admissible, 9
    feasible, 9
primal variable, 10
projection formula, 47

second-order sufficient condition, 11
security set, 25
set
    admissible, 9
    feasible, 9
set-valued mapping, 54
solution
    local optimal, 9
    strict local optimal, 9
    strict local optimal in the sense of L^∞, 18

space
    Banach, 90
    Lebesgue, 90
    Sobolev, 91
state, 6
strong regularity, 57

theorem
    implicit function, 57
    Sobolev's embedding, 93
    trace, 93
Tikhonov regularization, 7
trace, 93

weak derivative, 90
weak solution, 96
weakly lower semicontinuous, 87

Bibliography

[1] R. Adams. *Sobolev Spaces.* Academic Press, New York-London, 1975. Pure and Applied Mathematics, Vol. 65.

[2] J.-J. Alibert and J.-P. Raymond. Boundary control of semilinear elliptic equations with discontinuous leading coefficients and unbounded controls. *Numer. Funct. Anal. Optim.*, 18(3-4):235–250, 1997.

[3] W. Alt. The Lagrange-Newton method for infinite-dimensional optimization problems. *Numerical Functional Analysis and Optimization*, 11:201–224, 1990.

[4] W. Alt. *Nichtlineare Optimierung.* Vieweg Studium: Aufbaukurs Mathematik. [Vieweg Studies: Mathematics Course]. Friedr. Vieweg & Sohn, Braunschweig, 2002. Eine Einführung in Theorie, Verfahren und Anwendungen. [An introduction to theory, procedures and applications].

[5] W. Alt, R. Griesse, N. Metla, and A. Rösch. Lipschitz stability for elliptic optimal control problems with mixed control-state constraints. *submitted.*

[6] W. Alt and K. Malanowski. The Lagrange-Newton method for nonlinear optimal control problems. *Comput. Optim. Appl.*, 2(1):77–100, 1993.

[7] W. Alt and K. Malanowski. The Lagrange-Newton method for state constrained optimal control problems. *Comput. Optim. Appl.*, 4(3):217–239, 1995.

[8] Jöran Bergh and Jörgen Löfström. *Interpolation spaces. An introduction.* Springer-Verlag, Berlin, 1976. Grundlehren der Mathematischen Wissenschaften, No. 223.

[9] M. Bergounioux, K. Ito, and K. Kunisch. Primal-dual strategy for constrained optimal control problems. *SIAM J. Control Optim.*, 37(4):1176–1194, 1999.

[10] J. F. Bonnans and A. Shapiro. *Perturbation analysis of optimization problems.* Springer Series in Operations Research. Springer-Verlag, New York, 2000.

[11] D. Braess. *Finite elements*. Cambridge University Press, Cambridge, third edition, 2007. Theory, fast solvers, and applications in elasticity theory, Translated from the German by Larry L. Schumaker.

[12] H. Brezis. *Analyse fonctionnelle*. Collection Mathématiques Appliquées pour la Maîtrise. [Collection of Applied Mathematics for the Master's Degree]. Masson, Paris, 1983. Théorie et applications. [Theory and applications].

[13] E. Casas. Control of an elliptic problem with pointwise state constraints. *SIAM J. Control Optim.*, 24(6):1309–1318, 1986.

[14] E. Casas. Boundary control of semilinear elliptic equations with pointwise state constraints. *SIAM J. Control Optim.*, 31(4):993–1006, 1993.

[15] P. G. Ciarlet. *The finite element method for elliptic problems*. North-Holland Publishing Co., Amsterdam, 1978. Studies in Mathematics and its Applications, Vol. 4.

[16] J. E. Dennis, Jr. and R. B. Schnabel. *Numerical methods for unconstrained optimization and nonlinear equations*, volume 16 of *Classics in Applied Mathematics*. Society for Industrial and Applied Mathematics (SIAM), Philadelphia, PA, 1996. Corrected reprint of the 1983 original.

[17] A. V. Dmitruk. Maximum principle for the general optimal control problem with phase and regular mixed constraints. *Comput. Math. Model.*, 4(4):364–377, 1993.

[18] A. Dontchev. Implicit function theorems for generalized equations. *Mathematical Programming*, 70:91–106, 1995.

[19] L. C. Evans. *Partial differential equations*, volume 19 of *Graduate Studies in Mathematics*. American Mathematical Society, Providence, RI, 1998.

[20] R. Griesse. *Parametric Sensitivity Analysis for Control-Constrained Optimal Control Problems Governed by Systems of Parabolic Partial Differential Equations*. Dissertation. University of Bayreuth, Bayreuth, 2003.

[21] R. Griesse, N. Metla, and A. Rösch. Local quadratic convergence of SQP for elliptic optimal control problems with mixed control-state constraints. *submitted*.

[22] R. Griesse, N. Metla, and A. Rösch. Local quadratic convergence of SQP for elliptic optimal control problems with nonlinear mixed control-state constraints. *submitted*.

[23] P. Grisvard. *Elliptic Problems in Nonsmooth Domains*. Pitman, Boston, 1985.

[24] C. Grossmann and H.-G. Roos. *Numerical treatment of partial differential equations*. Universitext. Springer, Berlin, 2007. Translated and revised from the 3rd (2005) German edition by Martin Stynes.

[25] M. Heinkenschloss and F. Tröltzsch. Analysis of the Lagrange-SQP-Newton Method for the Control of a Phase-Field Equation. *Control Cybernet.*, 28:177–211, 1998.

[26] H. Heuser. *Funktionalanalysis*. Mathematische Leitfäden. [Mathematical Textbooks]. B. G. Teubner, Stuttgart, third edition, 1992. Theorie und Anwendung. [Theory and application].

[27] A. N. Kolmogorov and S. V. Fomīn. *Reelle Funktionen und Funktionalanalysis*. VEB Deutscher Verlag der Wissenschaften, Berlin, 1975. Übersetzt von der dritten russischen Auflage von D. Freitag, H. Palme und B. Stöckert, Hochschulbücher für Mathematik, Band 78.

[28] L. P. Lebedev and I. I. Vorovich. *Functional analysis in mechanics*. Springer Monographs in Mathematics. Springer-Verlag, New York, 2003. Revised and extended translation of the Russian original.

[29] J.-L. Lions. *Quelques méthodes de résolution des problèmes aux limites non linéaires*. Dunod, 1969.

[30] M. Mateos. *Problemas de control óptimo gobernados por ecuaciones semilineales con restrictiones de tipo integral sobre el gradiente del estado*. Dissertation. University of Cantabria, Santander, 2000.

[31] H. Maurer and J. Zowe. First and second order necessary and sufficient optimality conditions for infinite-dimensional programming problems. *Mathematical Programming*, 16:98–110, 1979.

[32] J. Nečas. *Les méthodes directes en théorie des équations elliptiques*. Masson et Cie, Éditeurs, Paris, 1967.

[33] P. Neittaanmaki, J. Sprekels, and D. Tiba. *Optimization of elliptic systems*. Springer Monographs in Mathematics. Springer, New York, 2006. Theory and applications.

[34] J. Nocedal and S. J. Wright. *Numerical optimization*. Springer Series in Operations Research. Springer-Verlag, New York, 1999.

[35] M. Renardy and R. C. Rogers. *An introduction to partial differential equations*, volume 13 of *Texts in Applied Mathematics*. Springer-Verlag, New York, second edition, 2004.

[36] S. Robinson. Strongly regular generalized equations. *Mathematics of Operations Research*, 5(1):43–62, 1980.

[37] A. Rösch and F. Tröltzsch. Sufficient second-order optimality conditions for a parabolic optimal control problem with pointwise control-state constraints. *SIAM Journal on Control and Optimization*, 42(1):138–154, 2003.

[38] A. Rösch and F. Tröltzsch. Existence of regular Lagrange multipliers for elliptic optimal control problems with pointwise control-state constraints. *SIAM Journal on Control and Optimization*, 45(2):548–564, 2006.

[39] A. Rösch and F. Tröltzsch. Sufficient second-order optimality conditions for an elliptic optimal control problem with pointwise control-state constraints. *SIAM Journal on Optimization*, 17(3):776–794, 2006.

[40] A. Rösch and F. Tröltzsch. On regularity of solutions and Lagrange multipliers of optimal control problems for semilinear equations with mixed pointwise control-state constraints. *SIAM J. Control Optim.*, 46(3):1098–1115 (electronic), 2007.

[41] H.-R. Schwarz. *Methode der finiten Elemente*, volume 47 of *Leitfäden der Angewandten Mathematik und Mechanik [Guides to Applied Mathematics and Mechanics]*. B. G. Teubner, Stuttgart, third edition, 1991. Eine Einführung unter besonderer Berücksichtigung der Rechenpraxis. [An introduction with special reference to computational practice], Teubner Studienbücher Mathematik. [Teubner Mathematical Textbooks].

[42] P. Spellucci. *Numerische Verfahren der nichtlinearen Optimierung*. Internationale Schriftenreihe zur Numerischen Mathematik. [International Series of Numerical Mathematics]. Birkhäuser Verlag, Basel, 1993.

[43] H. Triebel. *Interpolation theory, function spaces, differential operators*. VEB Deutscher Verlag der Wissenschaften, Berlin, 1978.

[44] F. Tröltzsch. On the Lagrange-Newton-SQP method for the optimal control of semilinear parabolic equations. *SIAM Journal on Control and Optimization*, 38(1):294–312, 1999.

[45] F. Tröltzsch. Lipschitz stability of solutions of linear-quadratic parabolic control problems with respect to perturbations. *Dynam. Contin. Discrete Impuls. Systems*, 7(2):289–306, 2000.

[46] F. Tröltzsch. *Optimale Steuerung partieller Differentialgleichungen. Theorie, Verfahren und Anwendungen.* Vieweg, Wiesbaden, 2005.

[47] F. Tröltzsch and S. Volkwein. The SQP method for control constrained optimal control of the Burgers equation. *ESAIM: Control, Optimisation and Calculus of Variations*, 6:649–674, 2001.

[48] K. Yosida. *Functional analysis.* Classics in Mathematics. Springer-Verlag, Berlin, 1995. Reprint of the sixth (1980) edition.

[49] K. Yosida and E. Hewitt. Finitely additive measures. *Trans. Amer. Math. Soc.*, 72:46–66, 1952.

[50] E. Zeidler. *Nonlinear Functional Analysis and its Applications.* Berlin, Springer-Verlag, 1990. Pure and Applied Mathematics, Vol. 65.

Südwestdeutscher Verlag für Hochschulschriften

Wissenschaftlicher Buchverlag bietet
kostenfreie
Publikation
von
Dissertationen und Habilitationen

Sie verfügen über eine wissenschaftliche Abschlußarbeit zu aktuellen oder zeitlosen Fragestellungen, die hohen inhaltlichen und formalen Anspruchen genügt, und haben **Interesse an einer honorarvergüteten Publikation?**

Dann senden Sie bitte erste Informationen über Ihre Arbeit per Email an: info@svh-verlag.de.

Unser Außenlektorat meldet sich umgehend bei Ihnen.

Südwestdeutscher Verlag für Hochschulschriften
Aktiengesellschaft & Co. KG

Dudweiler Landstr. 99
D – 66123 Saarbrücken
www.svh-verlag.de

Printed by Books on Demand GmbH, Norderstedt / Germany